# WJEC
# Mathematics
## for AS Level
## Pure & Applied
## Practice Tests

Stephen Doyle

Illuminate
Publishing

Published in 2018 by Illuminate Publishing Limited, an imprint of Hodder Education, an Hachette UK Company, Carmelite House, 50 Victoria Embankment, London EC4Y 0DZ

Orders: Please visit www.illuminatepublishing.com
or email sales@illuminatepublishing.com

British Library Cataloguing in Publication Data

A catalogue record for this book is available from the British Library

ISBN 978 1 911208 53 2

Printed by Ashford Colour Press, UK

11.22

The publisher's policy is to use papers that are natural, renewable and recyclable products made from wood grown in sustainable forests. The logging and manufacturing processes are expected to conform to the environmental regulations of the country of origin.

Editor: Geoff Tuttle
Cover design: Neil Sutton
Text design and layout: GreenGate Publishing Services, Tonbridge, Kent

**Photo credits**

**Cover:** Klavdiya Krinichnaya/Shutterstock; **p5** Radachynskyi Serhii/Shutterstock; **p7** Anatoli Styf/Shutterstock; **p13** Jeeraphun Kulpetjira/Shutterstock; **p19** Jan Miko/Shutterstock; **p22** Africa Studio/Shutterstock; **p26** Fotos593/ Shutterstock; **p29** jo Crebbin/Shutterstock; **p32** Jose Antonio Perez/ Shutterstock; **p38** Rena Schild/Shutterstock; **p44** Marco Ossino/Shutterstock; **p48** Sergey Nivens/Shutterstock; **p52** Blackregis/Shutterstock; **p56** urickung/ Shutterstock; **p62** Gajus/Shutterstock; **p66** cosma/Shutterstock; **p68** Will Rodrigues/Shutterstock; **p75** Andrea Danti/Shutterstock; **p79** sirtravelalot/Shutterstock; **p82** and **p93** panitanphoto/Shutterstock.

**Acknowledgements**

The author and publisher wish to thank Sam Hartburn for her help and careful attention when reviewing this book.

# Contents

## Questions

### AS Unit 1  Pure Mathematics A

### AS Unit 2  Applied Mathematics A

### Sample Test Papers

## Answers

# Introduction

This Practice Tests book is designed to be used alongside the WJEC Mathematics for AS Level Pure and Applied textbooks. The book follows the exact topic order in the books which is the same as the topic order in the WJEC specification.

The main purpose of this book is to build confidence in the topics by providing carefully graded questions, many of which are similar to the sorts of questions you might get in the actual examination.

Here are some of the features of the Practice Tests Book:

- Facts and formulae at the start of each topic making it easy for you to see what you should already know.

- Topic by topic carefully graded questions containing space for you to write your answers.

- Full answers at the back to all the questions with full explanations and explanations of alternative ways of solving the same problem.

- Tips on answering the questions to maximise your marks.

- Unstructured questions are provided which are a new feature of the specification.

- Specimen test papers for you to try.

# 1 Proof

## Essential facts and formulae

### Facts

There are the following available proofs to choose from (unless you are required to use a named proof in a question):

- **Proof by exhaustion** – uses all the allowable values to prove a mathematical statement true or false. Use only when there are a small number of possible values to try.

- **Disproof by counter-example** – uses an example to prove that the property does not work.

- **Proof by deduction** – uses something known or assumed (usually algebra) to decide whether a statement is true or false.

## Questions

**1** Prove that if $n$ is an integer between and including 1 and 7, then the expression $n^2 + 2$ is not a multiple of 4.   [2]

**2** Use disproof by counter-example to determine whether the statement 'if $x > y$, then $x^2 > y^2$, is true or false.   [2]

**3** In each of the two statements below, $c$ and $d$ are real numbers. One of the statements is true while the other is false.

    **A** Given that $(2c + 1)^2 = (2d + 1)^2$, then $c = d$.

    **B** Given that $(2c + 1)^3 = (2d + 1)^3$, then $c = d$.

(a) Identify the statement which is false. Find a counter-example to show that this statement is in fact false.

(b) Identify the statement which is true. Give a proof to show that this statement is in fact true.     [5]

**4** Prove by deduction that the sum of the squares of any two consecutive integers is always an odd number.     [2]

**5** If $n$ is an integer, prove by exhaustion that every cube number is either a multiple of 9 or is 1 more or 1 less than a multiple of 9.     [4]

# 2 Algebra and functions

## Essential facts and formulae

### Facts

**Rules of indices:** $a^m \times a^n = a^{m+n}, \quad a^m \div a^n = a^{m-n}, \quad (a^m)^n = a^{m \times n}, \quad a^{-m} = \dfrac{1}{a^m},$

$a^0 = 1$ (only if $a \neq 0$), $\quad a^{\frac{m}{n}} = \sqrt[n]{a^m}, \quad a^{-\frac{m}{n}} = \dfrac{1}{\sqrt[n]{a^m}}$

**Surds:** $\sqrt{a} \times \sqrt{a} = a, \quad \sqrt{a} \times \sqrt{b} = \sqrt{ab}, \quad (\sqrt{a} + \sqrt{b})(\sqrt{a} - \sqrt{b}) = a - b$

**Rationalisation:** $\dfrac{a}{b\sqrt{c}} = \dfrac{a}{b\sqrt{c}} \times \dfrac{\sqrt{c}}{\sqrt{c}} = \dfrac{a\sqrt{c}}{bc}$

$\dfrac{a}{\sqrt{b} \pm \sqrt{c}} = \dfrac{a}{(\sqrt{b} \pm \sqrt{c})} \times \dfrac{\sqrt{b} \mp \sqrt{c}}{\sqrt{b} \mp \sqrt{c}} = \dfrac{a\sqrt{b} \mp a\sqrt{c}}{b - c}$

Check back at the text book (Topic 2) for the following methods/techniques needed to answer questions in this topic:

- Completing the square
- The meaning of the discriminants of quadratic functions
- Sketching a quadratic function
- Solving linear inequalities
- Solving quadratic inequalities
- Transformations of the graph $y = f(x)$
- The remainder theorem
- The factor theorem

### Formulae

The quadratic equation $ax^2 + bx + c$ has solutions/roots given by

$$x = \dfrac{-b \pm \sqrt{b^2 - 4ac}}{2a}$$

The discriminant of a quadratic $= b^2 - 4ac$

# Questions

**1** Solve the following inequalities:

(a) $1 - 5x > -2x + 7$ [1]

(b) $\frac{x}{4} \le 2(1 - x)$ [2]

(c) $2x^2 + 5x - 12 \le 0$ [2]

**2** (a) Given that $x - 7$ is a factor of $x^3 - 8x^2 - px + 84$, write down an equation satisfied by $p$. Hence show that $p = 5$. [2]

(b) Solve the equation $x^3 - 8x^2 - 5x + 84 = 0$ [4]

**3** Solve the following inequalities:

(a) $1 - 2x < 4x + 7$ [2]

(b) $\frac{x}{2} \ge 2(1 - 3x)$ [2]

**4** The polynomial $f(x)$ is defined by
$$2x^3 + 7x^2 - 7x - 12$$
Find all the solutions of $f(x) = 0$ [3]

**5** (a) Expand and simplify $(4 - 2\sqrt{5})(3 + 4\sqrt{5})$ [2]

(b) Express $\sqrt{27} + \dfrac{81}{\sqrt{3}}$ in the form $a\sqrt{3}$, where $a$ is a constant. [2]

**6** Find the range of values for $m$ for which the equation $3x^2 + mx + 12 = 0$
does not have real roots. [4]

**7** (a) Express $\sqrt{27} + \sqrt{48}$ in the form $a\sqrt{3}$. [2]

(b) Express $\dfrac{20}{2 - \sqrt{2}}$ in the form $b + c\sqrt{d}$. [2]

**8** (a) Find the minimum value of $3x^2 - 12x + 10$ [3]

(b) Hence find the maximum value of $\dfrac{1}{3x^2 - 12x + 10}$. [1]

**9** Evaluate the following

(a) $\left(\dfrac{27}{9}\right)^0$ [1]

(b) $27^{\frac{2}{3}}$ [1]

(c) $\left(\dfrac{27}{8}\right)^{-\frac{1}{3}}$ [2]

**10** Figure 1 shows a sketch of the graph of $y = f(x)$. The graph has a minimum point at $(-3, -4)$ and intersects the $x$-axis at the points $(-8, 0)$ and $(2, 0)$.

(a) Sketch the graph of $y = f(x + 3)$, indicating the coordinates of the stationary point and the coordinates of the points of intersection of the graph with the $x$-axis. [3]

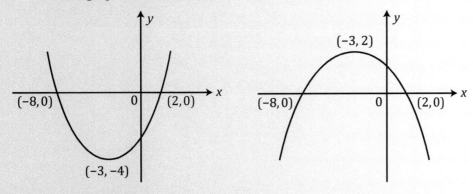

Figure 1                    Figure 2

(b) Figure 2 shows a sketch of the graph having one of the following equations with an appropriate value of either $p$, $q$ or $r$.

$y = f(px)$, where $p$ is a constant

$y = f(x) + q$, where $q$ is a constant

$y = rf(x)$, where $r$ is a constant

Write down the equation of the graph sketched in Figure 2, together with the value of the corresponding constant. [2]

**11** Sketch a graph of $y = \frac{1}{x}$ and hence use your graph to solve the
inequality $\frac{1}{x} < \frac{1}{2}$             [3]

**12**

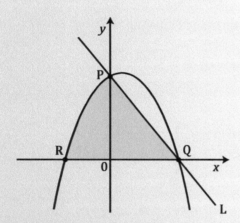

The diagram above shows a sketch of the curve $y = -x^2 + 2x + 8$. The curve
cuts the $x$-axis at points Q and R and the $y$-axis at point P. Line L passes
through P and Q.

Point T is a point that lies within the shaded area shown on the graph. Write
down three inequalities that must be satisfied by the coordinates of point T. [5]

# 3 Coordinate geometry in the (x, y) plane

## Essential facts and formulae

### Facts

- Straight line graphs are also called linear graphs and have an equation of the form $y = mx + c$

Notice there is a single $y$ on the left-hand side of the equation.

- $m$ is the gradient (i.e. the steepness of the line) and $c$ is the intercept on the $y$-axis.

- For two lines to be parallel to each other, they must have the same gradient.

- When two lines are perpendicular to each other (i.e. they make an angle of 90°), the product of their gradients is −1 so if one line has a gradient $m_1$ and the other a gradient of $m_2$ then $m_1 m_2 = -1$.

### Formulae

The **gradient** of the line joining points $(x_1, y_1)$ and $(x_2, y_2)$ is:

$$\text{Gradient} = \frac{y_2 - y_1}{x_2 - x_1}$$

The **mid-point** of a line joining the points $(x_1, y_1)$ and $(x_2, y_2)$ is given by:

$$\text{Mid-point} = \left( \frac{x_1 + x_2}{2}, \frac{y_1 + y_2}{2} \right)$$

The **equation of a straight line** with gradient $m$ and which passes through a point $(x_1, y_1)$ is given by:

$$y - y_1 = m(x - x_1)$$

The **length of a straight line** joining the two points $(x_1, y_1)$ and $(x_2, y_2)$ is given by:

$$\sqrt{(x_2 - x_1)^2 + (y_2 - y_1)^2}$$

The **equation of a circle** can be written in the form:

$$(x - a)^2 + (y - b)^2 = r^2, \text{ where the centre is } (a, b) \text{ and radius is } r.$$

An alternative form for the equation of a circle:

$$x^2 + y^2 + 2gx + 2fy + c = 0$$

where centre $(-g, -f)$ and radius given by $\sqrt{g^2 + f^2 - c}$.

# Questions

1. A straight line has the equation $2y = 4x - 5$
   (a) Write down the gradient of this line. [1]
   (b) A different line is drawn perpendicular to this line. Write down the gradient of the perpendicular line. [1]

2. A straight line passes through the points A(-2, 0) and B(6, 4).
   (a) Find the gradient of the line AB. [1]
   (b) The mid-point of AB is M. Find the coordinates of M. [1]
   (c) A straight line is drawn through point M which is perpendicular to the line AB.
   (i) Write down the gradient of this line. [1]
   (ii) Find the equation of this line. [1]

3. The line PQ has equation $2x + 3y = 5$. The point R has coordinates (3, 3).
   (a) Find the gradient of line PQ. [1]
   (b) A line is drawn which passes through R and is parallel to PQ. If this line crosses the y-axis at point S, find the coordinates of S. [2]

**4** The line AB has equation $4x + 5y = 10$.

(a) Find the gradient of AB. [1]

(b) Prove that the point C $(-5, 6)$ lies on AB. [1]

(c) Find the equation of the line through C which is at right-angles to AB. [2]

**5** Circle C has centre A and equation $x^2 + y^2 + 6x + 8y - 10 = 0$

Find the co-ordinates of A and find the radius of C. [2]

**6** P is the point $(0, 6)$ and Q is the point $(5, p)$.

(a) (i) Find the gradient of the line with equation $2x + 5y = 40$. [1]

(ii) Find the equation of the line through P which is parallel to the line $2x + 5y = 40$. [1]

(b) The line through P also passes through the point $(5, p)$. Find the value of $p$. [2]

**7** Find the equation of the circle having centre (1, 2) and passing through the point (4, −1). [3]

**8** (a) Find the coordinates of the centre of the circle with equation:

$x^2 + y^2 - 4x + 8y + 4 = 0$ [1]

(b) Prove that the point P (6, −4) lies on the circle. [1]

**9** A($-1$, 1), B(1, 2), C(4, 1) are points and P is a point such that AP is the diameter of a circle of centre B.

(a) Find the equation of the circle in the form $x^2 + y^2 + ax + by + c = 0$ where $a, b, c$ are constants to be found. [3]

(b) Prove that line CP is a tangent to the circle. [3]

**10** AB is the diameter of a circle with centre C and point P lies on the circumference.

If AP = $3\sqrt{5}$ and BP = $4\sqrt{5}$ :

(a) The diameter of the circle can be expressed as $k\sqrt{5}$. Find the value of $k$. [2]

(b) The centre of the circle C is at (3, $-2$). Find the equation of the circle giving your answer in the form $x^2 + y^2 + 2gx + 2fy + c = 0$ [3]

**11** Find the values of $m$ for which the line $y = m(x - 1)$ is a tangent to the curve $y = x^2 + 3$ [5]

**12** The line joining the points P(1, 3) and Q(3, −1) is a diameter of circle C with centre A and radius $r$.

(a) Find the equation of circle C in the form $x^2 + y^2 + ax + by + c = 0$ where $a$, $b$ and $c$ are constants to be determined. [3]

(b) The point R (1, −1) lies on the circle. Find the size of angle PQR giving your answer in degrees to one decimal place. [3]

# 4 Sequences and series – the binomial theorem

## Essential facts and formulae

### Facts

**Pascal's triangle**

```
            1
          1   1
        1   2   1
      1   3   3   1
    1   4   6   4   1
  1   5  10  10   5   1
```

### Formulae

**The binomial expansion of $(a + b)^n$ for positive integer $n$**

$$(a + b)^n = a^n + \binom{n}{1}a^{n-1}b + \binom{n}{2}a^{n-2}b^2 + \dots + \binom{n}{r}a^{n-r}b^r + \dots + b^n$$

$$\binom{n}{r} = {}^nC_r = \frac{n!}{r!(n-r)!}$$

**The binomial expansion of $(1 + x)^n$ for positive integer $n$**

$$(1 + x)^n = 1 + nx + \frac{n(n-1)}{2!}x^2 + \frac{n(n-1)(n-2)}{3!}x^3 + \dots$$

# Questions

**1** Use the binomial expansion to expand $(3 + 2x)^3$. [2]

**2** Find the term in $x^2$ in the binomial expansion of $\left(x + \dfrac{3}{x}\right)^6$. [2]

**3** (a) Expand $(1 + x)^7$ in ascending powers of $x$ up to, and including, the term in $x^3$. [3]

  (b) Using your result for part (a) find an approximation for $1.1^7$. Note that you must show all your working and simply putting $1.1^7$ into your calculator, will gain you no marks. [2]

  (c) Explain how the expansion could be used to find an approximate value for $0.99^7$. [1]

**4** (a) Expand $(1 + x)^6$ simplifying each term of the expansion. [3]

  (b) Use your expansion from part (a) to calculate the value of $(1.02)^6$ giving your answer to four decimal places. [2]

# 5 Trigonometry

## Essential facts and formulae

### Facts

You need to be able to derive the following using either right-angled or equilateral triangles.

$$\sin 30° = \frac{1}{2}$$

$$\cos 30° = \frac{\sqrt{3}}{2}$$

$$\tan 30° = \frac{1}{\sqrt{3}}$$

$$\sin 60° = \frac{\sqrt{3}}{2}$$

$$\cos 60° = \frac{1}{2}$$

$$\tan 60° = \sqrt{3}$$

### Formulae

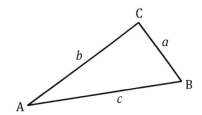

Sine rule: $\quad \dfrac{a}{\sin A} = \dfrac{b}{\sin B} = \dfrac{c}{\sin C} \quad$ or $\quad \dfrac{\sin A}{a} = \dfrac{\sin B}{b} = \dfrac{\sin C}{c}$

Cosine rule: $\quad a^2 = b^2 + c^2 - 2bc\cos A$

Area of triangle $= \dfrac{1}{2} ab \sin C$

### Trigonometric relationships

$$\tan \theta = \frac{\sin \theta}{\cos \theta}$$

$$\cos^2 \theta + \sin^2 \theta = 1$$

# Questions

**1** Find all the values of $\theta$ in the range $0° \leq \theta \leq 360°$ satisfying the equation
$(2\cos\theta - 1)(\cos\theta + 1) = 0$ [2]

**2** Find all the values of $\theta$ in the range $0° \leq \theta \leq 360°$ satisfying the equation
$3\cos^2\theta - \cos\theta - 2 = 0$ [4]

**3** Find all values of $\theta$ in the range $0° \leq x \leq 360°$ satisfying:

(a)  $3 \sin \theta = 1$  [2]

(b)  $\tan \theta = \dfrac{\sqrt{3}}{2}$  [2]

(c)  $3 \cos 2\theta = -1$  [2]

(d)  $2 \cos^2 \theta + \sin \theta - 1 = 0$  [2]

> Use either the symmetry of the graph or the CAST method to find the angles. Make sure you only include the angles in the range specified in the question.

**4** (a)  Find all values of $\theta$ in the range $0° \leq \theta \leq 360°$ satisfying
$$6 \sin^2 \theta + 1 = 2(\cos^2 \theta - \sin \theta).$$  [6]

(b)  Find all values of $x$ in the range $0° \leq x \leq 180°$ satisfying
$$\tan (3x - 57°) = -0.81.$$  [4]

(c)  Without carrying out any calculations, explain why there are no values of $\varphi$ which satisfy the equation
$$2 \sin \phi + 4 \cos \phi = -7.$$  [1]

**5** The diagram, right, shows a sketch of the triangle ABC with AB = $x$ cm, BC = $(x + 5)$ cm, AC = 7 cm and $\cos BAC = -\frac{3}{5}$.

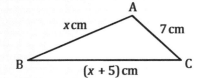

(a) Write down an equation satisfied by $x$. Hence show that $x = 15$. [3]

(b) Find the **exact** value of the area of triangle ABC. [3]

(c) The point D lies on BC and is such that AD is perpendicular to BC. Find the **length** of AD. [2]

**6** The triangle ABC has AB = 8 cm, BC = 15 cm and $A\hat{B}C = 60°$.

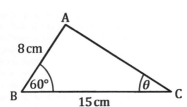

(a) Calculate the length of side AC. [2]

(b) Find the size of angle $\theta$ giving your answer to the nearest 0.1°. [3]

25

# 6 Exponentials and logarithms

## Essential facts and formulae

### Facts

A logarithm of a positive number to a base $a$ is the power to which the base must be raised in order to give the positive number.

$$y = a^x$$

$$\log_a y = x$$

For a positive base $a$, provided $a \neq 1$ the following are true:

$$\log_a a = 1, \text{ as } a^1 = a$$

$$\log_a 1 = 0, \text{ as } a^0 = 1$$

> You must remember both of these equations and be able to use them.

### Formulae

Rules of logarithms:

$$\log_a x + \log_a y = \log_a (xy)$$

$$\log_a x - \log_a y = \log_a \frac{x}{y}$$

$$\log_a x^k = k \log_a x$$

## Questions

**1** Show that $\log_3 \frac{1}{2^3} + \log_3 27 + 3 = 6 - \log_3 2$      [2]

**2** Simplify $\log_2 36 - \log_2 15 + \log_2 100 + 1$ [2]

**3** Express as a single logarithm in its simplest form:
$$3\log_{10} 4 - \frac{1}{2}\log_{10} 64 + 1$$ [3]

**4** If $a$ is a positive whole number, prove that for any value of $a$
$$\log_3 a \times \log_a 15 = \log_3 15$$ [3]

**5** Solve the equation $2^{3-2x} = 5$ giving the answer correct to two decimal places. [3]

**6** Solve the following equation giving the values of $x$ to 2 decimal places if not exact.
$$9^x - 5(3^x) + 6 = 0$$
[4]

# 7 Differentiation

## Essential facts and formulae

### Facts

To differentiate terms of a polynomial expression:

Multiply by the index and then reduce the index by one (i.e. if $y = kx^n$ then the derivative $\frac{dy}{dx} = nkx^{n-1}$).

To find if a function is increasing or decreasing at a certain point – differentiate the function and substitute the $x$-coordinate of the point into the expression. If it is positive, the function is increasing and if it is negative it is decreasing.

### Finding a stationary point

Put $\frac{dy}{dx} = 0$ and solve the resulting equation to find the value or values of $x$.

Substitute the value or values of $x$ into the equation of the curve to find the corresponding $y$-coordinate(s).

### Finding whether a stationary point is a maximum or minimum

Differentiate the first derivative (i.e. $\frac{dy}{dx}$) to find the second order derivative (i.e. $\frac{d^2y}{dx^2}$) at stationary points.

Substitute the $x$-coordinate of the stationary point into the expression for $\frac{d^2y}{dx^2}$ and look for the sign. If it is negative then the stationary point is a maximum point and if it is positive, then the stationary point is a minimum point.

If $\frac{d^2y}{dx^2} = 0$, then the result is inconclusive and further investigation is required.

### Curve sketching

Find the points of intersection with the $x$ and $y$-axes by putting $y = 0$ and $x = 0$ in turn and then solving the resulting equations.

Find the stationary points and their nature (i.e. maximum and minimum).

Plot the above on a set of axes.

### Formulae

If $y = kx^n$ then the derivative $\frac{dy}{dx} = nkx^{n-1}$

# Questions

**1** Given that $y = x^3 - 5x$, show from first principles that $\dfrac{dy}{dx} = 3x^2 - 5$      [5]

**2** Given that $y = \sqrt[3]{x^2} + \dfrac{64}{x}$, find the value of $\dfrac{dy}{dx}$ when $x = 8$.      [4]

**3**

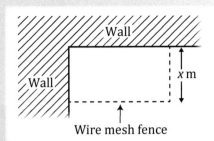

The diagram shown above shows two walls forming the sides of a sheep enclosure. A wire mesh fence is used to form the other two walls. The width of the enclosure is $x$ m. 25 m of wire mesh fence are used to form the other two walls.

(a) Show that the area, $A$ m$^2$, of the enclosure is given by $A = 25x - x^2$.  [1]

(b) (i) Find the value of $x$ that will make the area of the enclosure a maximum value.  [2]

    (ii) Find the maximum value of $A$.  [1]

**4** Find the range of values for which the function

$$f(x) = \frac{x^3}{3} - x^2 - 8x + 3 \qquad \text{is a decreasing function.} \qquad [5]$$

# 8  Integration

## Essential facts and formulae

### Facts

Students frequently lose **marks** because they differentiate instead of integrating. Another way they lose **marks** is forgetting to include the constant of integration.

Indefinite integration is the reverse process to differentiation. When integrating indefinitely you must always include the constant of integration, $c$.

$$\int x^n \, \mathrm{d}x = \frac{x^{n+1}}{n+1} + c \quad \text{(provided } n \neq -1)$$

$$\int k x^n \, \mathrm{d}x = \frac{k x^{n+1}}{n+1} + c \quad \text{(provided } n \neq -1)$$

Integrals in the form $\int_a^b y \, \mathrm{d}x$ are called definite integrals because the result will be a definite answer, usually a number, with no constant of integration.

### Definite integration as the area under a curve

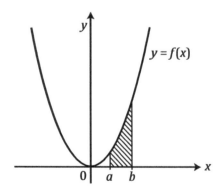

Shaded area $= \int_a^b y \, \mathrm{d}x$ where $a$ and $b$ are the limits.

A definite integral is positive for areas above the $x$-axis and negative for areas below the $x$-axis.

A final area must always be given as a positive value.

# Questions

**1** Find $\int(5x^4 + 4x^3 - 2x^2 + x - 1)dx$         [2]

**2** Find $\int(x-1)(x+8)dx$         [2]

**3** If $y = \int(3x^2 - 10x + 4)dx$, find the integral if it is known
that when $x = 2, y = 4$.         [3]

**4** Find $\int\left(\dfrac{x^2}{5} + \dfrac{x}{2}\right)dx$         [2]

**5** Find $\int_0^1 \frac{2}{3}(5x - 6)\,dx$. [3]

**6** (a) Sketch the curve with equation $y = x^2 - 4$ showing the points where the curve cuts the $x$-axis. [2]

(b) Find $\int_2^3 (x^2 - 4)\,dx$ and $\int_0^2 (x^2 - 4)\,dx$. [4]

(c) Explain why one of these integrals is positive and the other is negative. [1]

**7** (a) Find $= \int \left( \dfrac{3}{\sqrt[4]{x}} - 9x^{\frac{5}{2}} \right) dx$ [2]

(b) The region R is bounded by the curve $y = 2x^2 + \dfrac{6}{x^2}$, the $x$-axis and the lines $x = 1$, $x = 4$. Find the area of R. [5]

**8** The gradient function of a curve is $\frac{dy}{dx} = x^2 + 2x - 8$. The curve passes through the point P(3, 0).

(a) Show that the equation of the curve is $\frac{x^3}{3} + x^2 - 8x + 6$. [3]

(b) Find the coordinates of the stationary points of this curve. [3]

(c) Sketch the curve showing the stationary points and the intercept on the y-axis. [3]

**9**

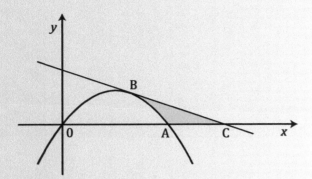

The diagram above shows a sketch of the curve $y = 3x - x^2$. The curve intersects the $x$-axis at the origin and at the point A. The tangent to the curve at the point B(2, 2) intersects the $x$-axis at the point C.

(a) Find the equation of the tangent to the curve at B. [4]

(b) Find the area of the shaded region. [8]

# 9  Vectors

## Essential facts and formulae

### Facts

Vectors have size and direction whereas scalars only have size.

### Formulae

#### Condition for two vectors to be parallel

For two vectors **a** and **b** to be parallel

$\quad$ **a** = $k$**b** $\qquad$ where $k$ is a scalar

#### The magnitude of a vector

The vector **r** = $a$**i** + $b$**j** has magnitude given by $|\mathbf{r}| = \sqrt{a^2 + b^2}$

#### The distance between two points

The distance between two points A$(x_1, y_1)$ and B$(x_2, y_2)$ is given by

$$d = \sqrt{(x_2 - x_1)^2 + (y_2 - y_1)^2}$$

#### The position vector of a point dividing a line in a given ratio

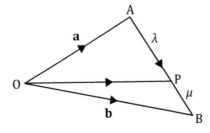

Point P dividing AB in the ratio $\lambda : \mu$ has position vector $\overrightarrow{\text{OP}}$ where

$$\overrightarrow{\text{OP}} = \frac{\mu\mathbf{a} + \lambda\mathbf{b}}{\lambda + \mu}$$

# Questions

**1** The vectors **a** and **b** are defined by $\mathbf{a} = 4\mathbf{i} - 3\mathbf{j}$, $\mathbf{b} = -2\mathbf{i} + 5\mathbf{j}$.

(a) Find the vector $2\mathbf{a} - \mathbf{b}$ [1]

(b) Vectors **a** and **b** are the position vectors of the points A and B respectively. Find the length of line AB. [3]

**2** The position vectors of the points A and B are given by

$$\mathbf{a} = \mathbf{i} + 3\mathbf{j} \quad \mathbf{b} = 3\mathbf{i} + 4\mathbf{j} \quad \text{respectively.}$$

(a) Write down the vector **AB**. [1]

(b) Find the exact magnitude of the vector **AB**. [2]

**3** The position vectors of the points A, B and C are given by

$$a = i + 2j \quad b = 2.5i + 3j \quad c = 4.5i \quad \text{respectively.}$$

(a) Determine whether or not the lines AB and BC are perpendicular, giving a reason for your answer. [2]

(b) Write down the vector **AC**. [2]

**4** The position vectors of the points A and B are given by

$$a = -2i - j \quad b = 4i + j \quad \text{respectively.}$$

(a) Write down the vector **AB**. [1]

(b) Points C and D have coordinates (3, 4) and (−3, 2) respectively. Prove that lines AB and DC are parallel and the same length. [3]

**5** Two vectors **a** and **b** are defined as follows:   $\mathbf{a} = 9\mathbf{i} - 2\mathbf{j}$   $\mathbf{b} = -3\mathbf{i} + 3\mathbf{j}$

(a) (i) Find the vector $2\mathbf{a} - 3\mathbf{b}$ [2]

(ii) The vectors **a** and **b** are the position vectors of the points P and Q respectively. Find the length of the line PQ. [2]

(b) Two ships T and U are a distance 10 km apart. The position vectors of ships T and U are denoted by **t** and **u** respectively.

(i) A lighthouse S lies on the line between T and U and a distance of 2 km from U. Find the position vector of S in terms of **t** and **u**. [2]

(ii) A rock has position vector $\frac{3}{5}\mathbf{u} + \frac{1}{5}\mathbf{t}$.

Explain why the rock cannot be on the straight line joining T and U. [3]

**6** A is the point with position vector **a** = −4**i** − 11**j** and B is the point with position vector **b** = 8**i** − 6**j**, both relative to the origin O.

   (a)  Find **AB** in terms of **i** and **j**.          [2]

   (b)  Find AB, the magnitude of vector **AB**.          [2]

   (c)  If **m** is the position vector of the mid-point of AB, find **m** in terms of **i** and **j**.          [3]

**7** A, B, C are three points given by position vectors **a** = 2**i** + 3**j**, **b** = 14**i** − 2**j**, **c** = −8**i** + 3**j** with respect to the origin O.

(a) Find **AB** in terms of **i** and **j**. [1]

(b) Find AB, the magnitude of **AB**. [2]

(c) Find the position vector of M, the midpoint of BC. [3]

(d) Find the position vector of the point P which divides AC in the ratio 3 : 7. [3]

# 1 Statistical sampling

## Essential facts and formulae

### Facts

**Population** – all members of the set that is being studied or has data collected about.

**Sample** – a smaller subset of the population and is used to draw conclusions about the population.

### Sampling techniques

**Simple random sampling** – each item in the population is given a number and then the required number of values in the sample is picked at random using calculator, program, website.

**Systematic sampling** – the sampling interval is found by dividing the population by the sample size. You then pick a random number within this sample size and start from that as the first item in the sample. You then add the sampling interval to the random number to get the next number in the sample. This is repeated until you have the required sample.

**Opportunity sampling** – here you decide on the sample size and simply use the most convenient way of collecting the sample (e.g. friends, relatives, classmates, work colleagues, etc.).

# Questions

1 A secondary school having 2000 students wishes to change the hours of the school day. The head teacher, before making a decision, would like to seek the views of the students.

   (a) Explain, in this context, the difference between a sample and a population. [2]

   (b) The head teacher wants to use a representative sample of 200 students. These students will be asked to complete a short questionnaire.

      (i) Explain how a simple random sample of 200 students could be obtained. [2]

      (ii) Give one advantage and one disadvantage of using simple random sampling. [2]

   (c) The deputy head suggests that systematic sampling could be used as an alternative to simple random sampling.

      (i) If 200 students are needed for the sample, work out the sampling interval. [1]

      (ii) Explain how the sampling interval is used to select the 200 students in the sample. [2]

      (iii) Give one advantage and one disadvantage of using systematic sampling. [2]

John wants to open a new coffee shop in an area where there aren't currently any. His friend owns a successful coffee shop in a different area and has offered to help John decide what to offer in terms of drinks and snacks. Their computer system records each sale and what was ordered, so they do a printout of 500 customers' orders. John decides to take a random sample of 50 customer orders to analyse.

Describe how John could obtain his random sample. [3]

**3** Being a vegan has become increasingly popular amongst young people. The biology teacher would like to know about their students' views on a vegan diet.

The teacher asked her 30 students on a scale of 1 to 5 how likely they would change to a vegan diet in the next two years. On the scale, 1 means 'highly unlikely' and 5 on the scale means 'certain'.

Here are her scores for the 30 students:

1, 1, 3, 1, 5, 4, 2, 4, 3, 1, 5, 1, 5, 3, 2, 1, 5, 3, 4, 4, 1, 2, 5, 3, 4, 5, 1, 1, 2, 3

(a) Taking an opportunity sample of the first 10 numbers in the list, calculate the mean score. [1]

(b) A systematic sample is to be taken of 5 data values.

   (i) Work out the sampling interval. [2]

   (ii) A random number was chosen in the sampling interval and it was 5. Using this value write down the list of data values in the sample. [2]

   (iii) Using the list from (ii), work out the mean score. [1]

(c) Explain the difference in the means obtained using the two sampling methods. [2]

(d) Compare the advantages and disadvantages of systematic sampling and opportunity sampling. [4]

# Data presentation and interpretation

## Essential facts and formulae

### Facts

#### Histograms

There are no gaps between the bars and the height of each bar is the frequency density.

The area of the bar is equal to the frequency, where:

Frequency (area of bar) = frequency density × class width

#### Box plots

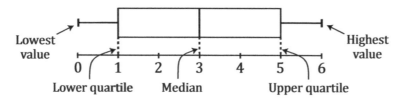

#### Calculation of the mean

For a **set of values**, mean, $\mu = \dfrac{\sum x_i}{n}$, where $\sum x_i$ is the sum of all the individual values and $n$ is the number of values

For a **frequency distribution**, mean, $\mu = \dfrac{\sum f x_i}{\sum f}$ , where $\sum f x_i$ is the sum of all the $x$ values, each multiplied by its frequency, $f$, and $\sum f$ is the sum of all the frequencies.

**Working out the mode** – the mode is the value(s) or class that occurs most often.

**Working out the median** – the median is the middle value when the data values are put in order of size. The median is at the $\dfrac{n + 1}{2}$ value.

#### *Measures of central variation*
#### *(variance, standard deviation, range and interquartile range)*

The following measures of the spread of data can be used:

#### Range

The difference between the largest value and the smallest value in a set of data.

## Interquartile range (IQR)

The difference between the upper quartile (Q3) and the lower quartile (Q1)
So IQR = Q3 − Q1

## Variance

Variance $= \dfrac{\sum (x_i - \mu)^2}{n}$ where $\sum (x_i - \mu)^2$ is the sum of the squares of the differences

between each value in the set and the mean $\mu$ and $n$ is the total number of values.

The **simplified formula for variance** is, Variance $= \dfrac{\sum x_i^2}{n} - \left(\dfrac{\sum x_i}{n}\right)^2$

where $\dfrac{\sum x_i^2}{n}$ is the mean of the squares of the values

and $\left(\dfrac{\sum x_i}{n}\right)^2$ is the square of the mean of the values.

## Standard deviation ($\sigma$)

The standard deviation ($\sigma$) is the square root of the variance so

$$\sigma = \sqrt{\dfrac{\sum (x_i - \mu)^2}{n}} \qquad \text{or} \qquad \sigma = \sqrt{\dfrac{\sum x_i^2}{n} - \left(\dfrac{\sum x_i}{n}\right)^2}$$

# Questions

**1** Here are some statements about data presentation and interpretation.
You have to decide whether each statement is true or false.

|  | | True | False | |
|---|---|:---:|:---:|---|
| (a) | Quantitative data is always numerical. | ☐ | ☐ | [1] |
| (b) | Discrete data can take all values. | ☐ | ☐ | [1] |
| (c) | Bar charts have quantitative data on the $x$-axis. | ☐ | ☐ | [1] |
| (d) | The height of a bar in a bar chart represents the frequency. | ☐ | ☐ | [1] |
| (e) | Histograms have gaps between the bars. | ☐ | ☐ | [1] |
| (f) | There are numerical values on both axes of a histogram. | ☐ | ☐ | [1] |
| (g) | Histograms always have bars of unequal width. | ☐ | ☐ | [1] |
| (h) | The area of a bar on a histogram represents the frequency density. | ☐ | ☐ | [1] |

**2** Here are some statements about data presentation and interpretation.
You have to decide whether each statement is true or false.

|  | | True | False | |
|---|---|:---:|:---:|---|
| (a) | A positively skewed distribution is skewed to the right. | ☐ | ☐ | [1] |
| (b) | A perfectly symmetrical distribution has no skew. | ☐ | ☐ | [1] |
| (c) | A scatter graph with negative correlation has a positive gradient. | ☐ | ☐ | [1] |
| (d) | The mode, mean and standard deviation are all measures of central tendency. | ☐ | ☐ | [1] |
| (e) | The interquartile range is the spread of the middle half of the data when the data is arranged in order of size. | ☐ | ☐ | [1] |

**3** A researcher collected data about the height and arm span in cm of 41 people.

(a) The pairs of values were used to produce a scatter graph (right) of 'respondent height in cm' against 'respondent arm span in cm'.

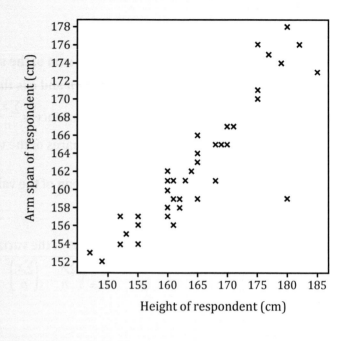

(i) Comment about the correlation between 'respondent height' and 'respondent arm span'. [1]

(ii) Interpret the correlation between respondent height' and 'respondent arm span' in this context. [1]

(b) The regression equation for the set of data used to produce the scatter graph is:

'Height of respondent' = 54.8 − 0.654 × 'Arm span of respondent'

(i) Interpret the gradient of the regression equation for this model. [1]

(ii) One of the points is an outlier. Circle the outlier on the graph and explain the effect removing this outlier would have on the mean respondent height and the mean respondent arm span. [3]

(iii) State, with a reason, whether the regression model would be useful to predict the height for a baby with arm span 20 cm. [1]

(iv) State whether the relationship between 'respondent height' and 'respondent arm span' is causal for an adult. Explain your answer. [1]

**4** Here are the marks in a maths exam for 10 boys and 10 girls:

Girls marks:  20, 41, 72, 24, 55, 40, 63, 85, 60, 62

Boys marks:  61, 40, 15, 21, 90, 62, 34, 40, 60, 53

(a) The following table has been partially completed with summary statistics for both sets of data. Complete this table.

| Summary statistic | Boys | Girls |
|---|---|---|
| Lower quartile | | 36 |
| Median | | 57.5 |
| Upper quartile | | 65.25 |
| Highest mark | | 85 |
| Lowest mark | | 20 |
| Mean | | 52.2 |
| Range | | 65 |
| Inter-quartile range | | 29.25 |

[5]

(b) The box and whisker diagram (see right) has been produced for the girls using some of the summary statistics.

On the graph paper below, draw a box and whisker diagram for the boys' marks.

[2]

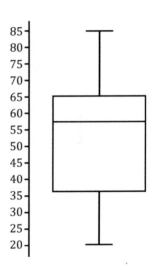

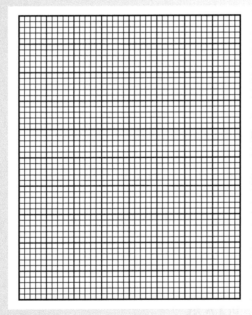

(c) Compare and contrast the two sets of data. [3]

# 3 Probability

## Essential facts and formulae

### Facts

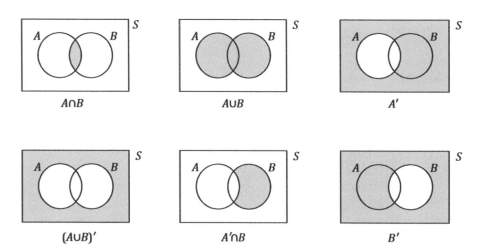

$A \cap B$

$A \cup B$

$A'$

$(A \cup B)'$

$A' \cap B$

$B'$

**Independent events** – where one event happening does not alter the probability of another event happening.

**Dependent events** – where one event happening alters the probability of another event happening.

**Mutually exclusive events** – events where one or other event can happen, but not both.

### Formulae

Generalised addition law: $P(A \cup B) = P(A) + P(B) - P(A \cap B)$

Addition law for mutually exclusive events: $P(A \cup B) = P(A) + P(B)$

Multiplication law for independent events: $P(A \cap B) = P(A) \times P(B)$

# Questions

**1** Two independent events $A$ and $B$ are such that $P(A) = 0.5$, $P(B) = 0.2$ and $P(A \cup B) = 6 \times P(A \cap B)$.

(a) Explain, by giving an example of each, the difference between mutually exclusive events and independent events. [4]

(b) Find $P(A \cap B)$. [1]

(c) Find $P(A \cup B)$. [1]

**2** The events $A$ and $B$ are such that $P(A) = 0.25$, $P(A \cup B) = 0.4$.

Evaluate $P(B)$ when:

(a) $A$, $B$ are mutually exclusive. [2]

(b) $A$, $B$ are independent. [3]

**3** The events $A$ and $B$ are such that
$P(A) = 0.2$, $P(B) = 0.5$, $P(A \cup B) = 0.65$.
(a) Determine whether or not $A$ and $B$ are independent. [3]
(b) Explain the significance of the result. [1]

**4** Two events $A$ and $B$ are such that $P(A) = 0.35$ and $P(B) = 0.45$.
Find the value of $P(A \cup B)$ when
(a) $A$, $B$ are mutually exclusive [2]
(b) $A$, $B$ are independent [2]
(c) $A \subset B$ [1]

**5** The independent events $A$, $B$ are such that $P(A) = 0.2$, $P(A \cup B) = 0.5$.

(a) Determine the value of $P(B)$. [4]

(b) Calculate the probability that exactly one of the events $A$, $B$ occurs. [3]

# 4 Statistical distributions

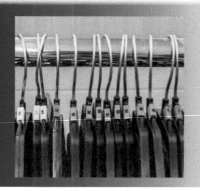

## Essential facts and formulae

### Facts

**Use the binomial distribution when there are:**

- Independent trials
- Trials where there is a constant probability of success
- A fixed number of trials (i.e. $n$ is known)
- Only success or failure

**Use the Poisson distribution to approximate the binomial distribution when:**

- $n$ is large (usually > 50)
- $p$ is small (usually < 0.1)

### The discrete uniform distribution

This is a distribution where all the outcomes are equally likely.
So if there are $N$ possible outcomes, the probability of a particular outcome $= \frac{1}{N}$.

### Formulae

#### The binomial distribution

For a fixed number of trials, $n$, each with a probability $p$ of occurring, the probability of a number of $x$ successes is given by:

$$P(X=x) = \binom{n}{x} p^x (1-p)^{n-x}$$

If $X$ is B$(n, p)$ then the mean $= np$

#### The Poisson distribution

In a particular interval, the probability of an event $X$ occurring $x$ number of times is given by the following formula:

$$P(X=x) = e^{-\lambda}\frac{\lambda^x}{x!} \qquad \text{where mean, } \mu = \lambda.$$

# Questions

**1** The binomial distribution can be used to model different situations. Here are some models. You have to decide, giving a reason, whether the binomial distribution is a suitable model or not.

(a) Modelling the number of throws of a die until a six is thrown. [1]

(b) Modelling of picking different numbers of red balls, when the balls being picked are not put back into the bag. Initially there are 6 red and 4 black balls in the bag. [1]

(c) Modelling the numbers of faulty light bulbs in a batch of 20 bulbs if it is known that the probability of selecting a faulty bulb is 0.03. [1]

(d) Modelling the number of times the bull is hit in a game of darts when 40 darts are thrown at the bull, if the probability of hitting the bull is 0.12. [1]

(e) Modelling the number of breakages in a batch of 50 wine glasses being delivered by a courier. [1]

(f) Modelling the number of snowdrop bulbs flowering out of 15 bulbs planted if the individual probability that a bulb flowers is 0.7. [1]

**2** The probability of being left-handed is 0.09. Find the probability that in a random sample of ten people:

(a) Exactly half of them are left-handed. [2]

(b) None of them are left-handed. [1]

(c) More than 5 of them are left-handed. [2]

(d) State one assumption you have made in all your answers. [1]

**3** When snowdrop bulbs are planted in autumn, there is a probability of each bulb growing of 0.65 independent of other snowdrop bulbs.

(a) If a clump of 30 bulbs are planted in autumn, find the probability that

   (i) exactly 20 of them grow. [2]

   (ii) at least 15 of them grow. [2]

(b) $n$ bulbs are planted in autumn and the probability that they all grow is 0.005688.

   Find the value of $n$. [4]

**4** (a) The random variable $X$ has the binomial distribution B(25, 0.8).

    (i) Without the use of tables, calculate $P(X = 10)$ giving your answer as a decimal to 3 significant figures,

    (ii) Determine $P(10 \leq X \leq 15)$.          [5]

(b) The random variable $Y$ has the binomial distribution B(300, 0.04).

    Use the Poisson distribution to determine the approximate value of $P(Y = 5)$.          [3]

**5** When Jack types a page of a document, the number of errors made may be modelled by a Poisson distribution with mean 0.8 . He types a 10-page document. Determine the probability that the total number of errors is less than 5.          [3]

**6** In a factory, the number of components rejected in any interval of time $t$ hours has a Poisson distribution with mean $0.1t$.

**Without the use of tables**, find the probability that the number of components rejected in the first 15 hours the machine is switched on is

(a) 2, [3]

(b) more than 2. [3]

**7** A machine mixes dough and raisins to make cookies. A batch of this dough is thoroughly mixed and then sliced up to produce cookies. One hundred cookies were made and they contained a total of 400 raisins. It is assumed that the raisins are randomly distributed over all the cookies.

(a) What is the probability that despite all the precautions, one or more cookies from the batch contained no raisins at all? [2]

(b) How many raisins should be put in the batch of dough to be 99% sure that no cookie comes out without any raisins in it? [4]

**8** A small office handling insurance claims gets on average 2.5 phone calls between 12 pm and 1 pm on weekdays. The office manager knows from experience that the staff available during this time can handle up to 5 calls per hour. Just to check, the manager would like to find the probability that 6 calls will be received during a particular day. Find this probability. [3]

9 A zoologist is studying a certain breed of dog.

(a) He knows from past experience that the probability of a newly born puppy being female is 0.55. He selects a random sample of 20 newly born puppies. Calculate the probability that the number of females in the sample is:

(i) exactly 12,

(ii) between 8 and 16 (both inclusive). [8]

(b) The probability of a newly born puppy being yellow is 0·05. Use an approximating distribution to find the probability that less than 5 out of a random sample of 60 newly born puppies are yellow. [3]

10 The number of emergency patients with toothache who need to be seen at a dental practice each day can be modelled by a Poisson distribution with mean 8.

Determine the probability that, on a randomly chosen day, the number of emergencies to be seen is:

(a) exactly 7 [2]

(b) fewer than 10. [2]

# 5 Statistical hypothesis testing

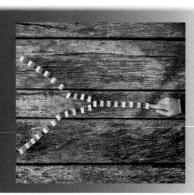

## Essential facts and formulae

### Facts

#### Hypothesis testing

Hypothesis testing is used to test a hypothesis about the probability of the number of times ($X$) a certain property crops up.

The test statistic $X$ is modelled by a binomial distribution $B(n, p)$ where $p$ is the probability of the event occurring in one trial and $n$ is the total number of trials.

#### Null and alternative hypothesis

Null hypothesis $\mathbf{H}_0 : p$ = a value

Alternative hypothesis for a one-tailed test is $\mathbf{H}_1 : p <$ a value  or $\mathbf{H}_1 : p >$ a value

Alternative hypothesis for a two-tailed test is $\mathbf{H}_1 : p \neq$ a value

#### Significance level

Assuming the null hypothesis is true, the significance level indicates how unlikely a value needs to be before the null hypothesis $\mathbf{H}_0$ is rejected.

The significance level for this topic can be 1% ($\alpha = 0.01$), 5% ($\alpha = 0.05$) or 10% ($\alpha = 0.1$).

#### The critical value and critical region

Here you use the significance level along with the values of $n$ and $p$ to find the set of values of $X$ that would cause the null hypothesis to be rejected.

In all of the following cases the significance level is taken as 5% but this can be changed to 1% or 10% or any other significance level.

If $\mathbf{H}_0$ contains a < sign the critical region will be the lower 5% of values of $X$ (i.e. in the lower tail) so use the tables and find the first $p$-value in the column which exceeds 0.05 and choose the value of $X$ **before** it. This is the critical value, $a$, and the critical region will be $X \leq a$.

If $\mathbf{H}_0$ contains a > sign the critical region will be the upper 5% of values of $X$ (i.e. in the upper tail) so use the tables and find the first $p$-value in the column which exceeds 0.05 and choose the value of $X$ **after** it. This is the critical value, $a$, and the critical region will be $X \geq a$.

If $H_0$ contains a ≠ sign the critical region will be the upper 2.5% and lower 2.5% of values of $X$ (i.e. in the upper tail and lower tail). We use the techniques outlined in the previous two paragraphs to find the two critical values and the critical regions in each of the two tails.

### Type I and type II errors

**Type I error** – A type I error is made when you incorrectly reject a true null hypothesis.

**Type II error** – A type II error is made when you fail to reject a false null hypothesis.

### *p*-values

The *p*-value is the probability that the observed result or a more extreme one will occur under the null hypothesis.

Assuming that the probability distribution of $X$ is B$(n, p)$ the probability of obtaining a value (called the *p*-value) of the test statistic or a more extreme one can be found using tables. If this value is less than or equal to the level of significance, the null hypothesis is rejected.

# Questions

1. A footballer who does not like taking penalties, tells his manager that he only has a 50% chance of scoring from a penalty. The manager thinks he is underestimating his ability at taking penalties and he is much better than this. The manager decides to test this during training and in 20 penalties, the footballer scores 15 times.

   Using a 5% level of significance, test the footballer's claim. [5]

**2** A coin is tossed 10 times and 7 heads are produced. Determine, at the 10% level of significance, whether or not the coin is fair. [5]

**3** Amy sees one or more red squirrels in her garden. She estimates the probability of seeing one or more red squirrels in her garden per day is 0.05.

This spring, she thinks this probability is different. She sees one or more red squirrels in her garden on 3 days out of 6.

Use a 1% level of significance to test her original probability. [6]

**4** A student claims he has a 45% chance of passing an exam. His friend says he hasn't worked hard enough and is therefore overestimating his chance of passing.

He sits 10 exams and passes $N$. Given that in a test, at the 5% level of significance, it was concluded that he has overestimated his chances of passing, what are the possible values of $N$? [5]

**5** In a dental practice the probability that patients have to wait more than 30 minutes after their appointment time is 0.3. Following changes in working practices, the lead dentist says there is a decrease in the numbers of patients having to wait more than 30 minutes. That day, she records the waiting times for the next 30 patients and 6 wait more than 30 minutes.

(a) (i)   Write down a suitable null hypothesis for the test. [1]

    (ii)  Write down a suitable alternative hypothesis for the test. [1]

    (iii) Write down a test statistic that can be used for the test. [1]

(b) Find the critical region if the significance level is 5%. [5]

(c) Comment on the lead dentist's claim. [2]

# 6 Quantities and units in mechanics

## Essential facts and formulae

### Facts

The fundamental quantities and their units are:

Length (m)

Time (s)

Mass (kg)

### Formulae

Derived quantities are quantities that are derived from the fundamental quantities using a formula and include the following:

$$\text{Velocity} = \frac{\text{length or distance}}{\text{time}} \ (\text{m s}^{-1})$$

$$\text{Acceleration} = \frac{\text{change in velocity}}{\text{time taken}} \ (\text{m s}^{-2})$$

$$\text{Force} = \text{mass} \times \text{acceleration} \ (\text{N})$$

# Questions

**1** Give the units of each of the following quantities.
   (a) Acceleration     (d) Weight
   (b) Velocity         (e) Force
   (c) Density          (f) Moment                    [3]

**2** The density of gold is 19.32 g cm$^{-3}$. Find this density in the units kg m$^{-3}$.     [2]

# 7 Kinematics

## Essential facts and formulae

### Facts

Displacement/distance–time graphs – the gradient is the velocity/speed.

Velocity/speed–time graphs –   the gradient is the acceleration or deceleration.
the area under the graph is the displacement/
distance travelled.

### Formulae

These formulae are used **throughout** the topic. You must remember them as they
will not be given in the exam. Remember that these equations can only be used
when the acceleration is constant.

$$\text{Speed} = \frac{\text{distance travelled}}{\text{time taken}}$$

$v = u + at$

$s = ut + \frac{1}{2}at^2$

$v^2 = u^2 + 2as$

$s = \frac{1}{2}(u + v)t$

$s$ = displacement
$u$ = initial velocity
$v$ = final velocity
$a$ = acceleration
$t$ = time

When the acceleration is **not** constant the following formulae are used:

Displacement ($r$) → *Differentiate* $\dfrac{dr}{dt}$ → Velocity ($v$) → *Differentiate* $\dfrac{dv}{dt}$ → Acceleration ($a$)

Displacement ($r$) ← $r = \int v\,dt$ ← *Integrate* ← Velocity ($v$) ← $v = \int a\,dt$ ← *Integrate* ← Acceleration ($a$)

# Questions

**1** A car starting from rest accelerates with uniform acceleration 0.9 m s$^{-2}$ in a straight line for 5 seconds.

It then maintains a constant speed for 20 seconds before being uniformly decelerated for 8 seconds before coming to rest.

(a) Sketch a velocity–time graph for the motion of the car. [2]

(b) Find the maximum velocity reached by the car. [1]

(c) Find the deceleration of the car. [1]

(d) Find the total distance travelled by the car. [1]

**2** A small object, of mass 0.02 kg, is dropped from rest from the top of a building which is 160 m high.

(a) Calculate the speed of the object as it hits the ground. [3]

(b) Determine the time taken for the object to reach the ground. [3]

(c) State one assumption you have made in your solution. [1]

**3** A particle is projected vertically upwards with a speed of $15\,\text{m s}^{-1}$.

    (a) Find the time in seconds for the particle to reach its greatest height.    [2]

    (b) Find the maximum height reached by the particle.    [2]

    (c) State one modelling assumption you have made in your solutions.    [1]

**4** The velocity–time graph, right, shows the motion of a particle in a straight line with constant acceleration. The particle passes the origin at $t = 0\,\text{s}$ with a velocity of $15\,\text{m s}^{-1}$.

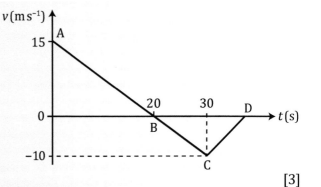

    (a) State what each of the following sections of the graph represents.

       (i) AB

       (ii) BC

       (iii) CD    [3]

    (b) If the total displacement shown by the graph is zero, find the time in seconds when the particle returns to the origin.    [3]

    (c) Find the average speed for the whole journey.    [1]

5  A particle passes the origin O with a velocity of $4 \text{ m s}^{-1}$ and accelerates along the positive $x$-axis with an acceleration of $4 \text{ m s}^{-2}$. Five seconds later a second particle sets out from rest from O and travels along the positive $x$-axis with an acceleration of $10 \text{ m s}^{-2}$. Find how far from O the second particle overtakes the first particle, giving your answer to the nearest metre.   [7]

**6** A stone is projected vertically from the ground with a speed of $14.7 \text{ m s}^{-1}$ from the edge of a cliff which is 49 m above the sea. The stone goes to its greatest height and then falls into the sea.

(a) Calculate the time the stone spends in the air. [3]

(b) Calculate the speed of the stone when it hits the sea. [3]

**7** A particle sets out from the origin O at $t = 0$ s.
It then moves along a horizontal axis so that its velocity $v$ m s$^{-1}$ $t$ seconds later is given by $v = 4 + 3t - t^2$.

(a) Calculate the time at which the particle is at rest and calculate the acceleration of the particle at this time. [2]

(b) Calculate the average velocity of the particle during the first four seconds after leaving O. [3]

**8** The speed of a particle in m s$^{-1}$ is given by $v = 8 + 7t - t^2$

(a) Find the time when the velocity is zero. [2]

(b) Find the velocity when $t = 0$ s. [1]

(c) Calculate the distance travelled by the particle in the 2nd second.
Give your answer as a mixed number. [3]

**9** A particle P, of mass 3 kg, moves along the horizontal $x$-axis under the action of a resultant force $F$ N. Its velocity $v$ m s$^{-1}$ at time $t$ seconds is given by:

$$v = 12t - 3t^2$$

(a) Given that the particle is at the origin O when $t = 1$, find an expression for the displacement of the particle from O at time $t$ s.                   [3]

(b) Find an expression for the acceleration of the particle at time $t$ s.         [2]

**10** A pebble is projected vertically upwards with a speed of 7 m s$^{-1}$ from the top of a cliff. It hits the ground at the bottom of the cliff 4 seconds later.

Determine the height of the cliff.                                            [5]

# 8 Dynamics of a particle

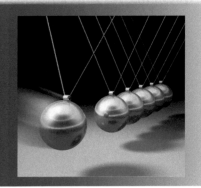

## Essential facts and formulae

### Facts

**Newton's 2nd law of motion** – unbalanced forces produce an acceleration according to the equation:

Force = mass × acceleration    or    $F = ma$  for short.

### Lifts accelerating, decelerating and travelling with constant velocity

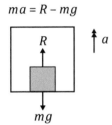

$ma = R - mg$

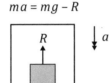

$ma = mg - R$

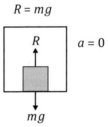

$R = mg$

### The motion of particles connected by strings passing over fixed pulleys or pegs

Pulleys or pegs are smooth so no frictional forces act.
Strings are light and inextensible so the tension remains constant.

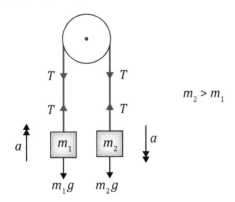

$m_2 > m_1$

Acceleration of each mass is the same as the string is taut.
Newton's second law can be applied to each mass separately.

For $m_1$,           $m_1 a = T - m_1 g$
while for $m_2$,     $m_2 a = m_2 g - T$

### Formulae

Force = mass × acceleration or $F = ma$ for short.

# Questions

 A girl of mass 58 kg stands on the floor of a lift which is descending with an acceleration of 2.5 m s$^{-2}$. Calculate the magnitude of the reaction of the floor of the lift on the girl. [3]

2 A lift starts from rest and travels upwards with a uniform acceleration of 4 m s$^{-2}$ until a constant speed of 12 m s$^{-1}$ is reached. The lift then travels with constant speed of 12 m s$^{-1}$ for 5 s and is then brought to rest in 4 s.

(a) Draw a velocity–time graph representing the motion of the lift and use the graph to find the deceleration of the lift. [3]

(b) A man of mass 50 kg stands in the lift during the above motion.

Calculate the magnitude of the reaction of the floor of the lift on the man during each of the three stages of its motion. [5]

3 The diagram shows a body A, of mass 5 kg, lying on a smooth horizontal table. It is connected to another body B, of mass 9 kg, by a light inextensible string, which passes over a smooth light pulley P fixed at the edge of the table so that B hangs freely.

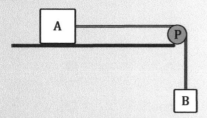

Initially, the system is held at rest with the string taut. A horizontal force of magnitude 126 N is then applied to A in the direction PA so that B is raised. Find the magnitude of the acceleration of A and the tension in the string. [7]

**4** Two particles A and B with masses $m$ kg and 5 kg respectively where $m < 5$, are connected by a light inextensible string which passes over a smooth light pulley. The arrangement is released from rest with both masses at the same height above a horizontal surface.

If the acceleration of the masses is $0.2g$,

(a) Show that $T = 4g$. [3]

(b) Find the size of mass, $m$ in kg giving your answer to 2 significant figures. [3]

(c) Explain how your answer has taken into account that the string is light inextensible and that the pulley is smooth. [2]

# 9   Vectors

## Essential facts and formulae

### Facts

Scalar quantities have magnitude (i.e. size) only and include distance and speed.

Vector quantities have both magnitude and direction and include displacement, velocity, acceleration and force.

Vectors are typed in bold and not in italics, so **s**, **r**, **v**, **a** and **F** are all vectors.

The resultant of vectors acting at a point can be found by adding the individual vectors.

### The magnitude of a vector

The vector $\mathbf{r} = a\mathbf{i} + b\mathbf{j}$ has magnitude given by $|\mathbf{r}| = \sqrt{a^2 + b^2}$

### The direction of a vector

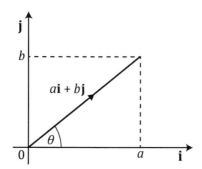

The angle made by the vector $a\mathbf{i} + b\mathbf{j}$ to the unit vector **i** is $\theta$, where $\theta = \tan^{-1}\left(\dfrac{b}{a}\right)$

### Converting from magnitude and direction to a vector

If you know the magnitude and direction of a vector you can convert this to vector form in the following way

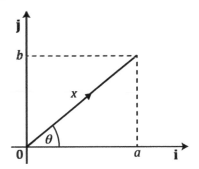

If the length of the vector is $x$ and it is inclined at an angle $\theta$ to the **i**-direction (or positive $x$-axis) then using trigonometry

$$\frac{a}{x} = \cos\theta°, \text{ so } a = x\cos\theta° \quad \text{and} \quad \frac{b}{x} = \sin\theta°, \text{ so } b = x\sin\theta°$$

Vector is $\mathbf{x} = a\mathbf{i} + b\mathbf{j}$

### Formulae

The vector $\mathbf{r} = a\mathbf{i} + b\mathbf{j}$ has magnitude given by $|\mathbf{r}| = \sqrt{a^2 + b^2}$

## Questions

**1** Here are some quantities. You have to decide whether they are scalar or vector quantities by placing a tick in the appropriate column.

| Quantity | Scalar | Vector |
|---|---|---|
| Velocity | | |
| Speed | | |
| Force | | |
| Distance | | |
| Acceleration | | |
| Displacement | | |

[6]

2  A particle of mass 5 kg has two forces **P** and **S** applied to it where:

$$P = 3i - 20j$$
$$S = -8i + 8j$$

Find

(a)  The resultant force **R**.                                                      [1]

(b)  The magnitude of the resultant force.                                [2]

(c)  The angle the resultant force makes to the **i** vector.        [2]

(d)  The acceleration **a**.                                                         [1]

3  A tractor travels from P to Q and then from Q to R.

The displacement from P to Q is $5i + 2j$ km.

The displacement from Q to R is $-3i + 3j$ km.

(a)  Find the total distance travelled from P to R                    [3]

(b)  Find the angle $\overrightarrow{PR}$ makes with the unit vector **i** giving your answer to
     one decimal place.                                                              [2]

1  (a)  Expand $\left(x - \dfrac{1}{x}\right)^4$.  [3]

(b)  Explain why the substitution $x = 1$ will help you to check your answer.  [1]

2  Show that $\dfrac{7}{2\sqrt{14}} + \left(\dfrac{\sqrt{14}}{2}\right)^3$ can be expressed as $k\sqrt{14}$ where $k$ is an integer.  [3]

**3** Using disproof by counter-example prove that the following statement is false. 'If two functions $f$ and $g$ are such that their derivatives $f'$ and $g'$ are equal, then the functions $f$ and $g$ must themselves be equal.'                                                [3]

**4** (a) Given that $x = \log_a y$,

     (i)   express $y$ in terms of $x$ and $a$,                              [1]

     (ii)  express $a$ in terms of $x$ and $y$.                              [1]

     Given also that $x = 2$, find the values of

     (iii) $\log_a y^3$                                                        [1]

     (iv) $\log_a (ay)^3$                                                     [1]

     (v) $\log_a\left(\dfrac{y^5}{a^4}\right)$                                 [1]

  (b) If $2^{x-1} = 3^{(x+3)}$, show that $x = \dfrac{3\log_{10} 3 + \log_{10} 2}{\log_{10} 2 - \log_{10} 3}$          [3]

*Questions* Sample Test Paper Unit 1

**5** Circle C has centre P and equation $x^2 + y^2 - 18x - 22y + 177 = 0$
   (a) Find the coordinates of P and the radius of the circle. [2]
   (b) (i) Prove that the point T(5, 8) lies on the circle. [1]
       (ii) Find the equation of the tangent to the circle C at point T. [4]

**6** Find the range of values for which the function
$$f(x) = \frac{x^3}{3} + \frac{x^2}{2} - 12x + 1$$
is an increasing function. [5]

**7** A plot of land is in the shape of a triangle ABC.

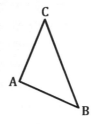

Given that $\overrightarrow{AB} = 240\mathbf{i} - 60\mathbf{j}$ metres and $\overrightarrow{BC} = -180\mathbf{i} + 200\mathbf{j}$ metres.

(a) Find $\overrightarrow{AC}$. [1]

(b) Find angle BAC giving your answer to the nearest 0.1°. [2]

(c) Find the area of the field in m² to the nearest integer. [2]

**8** Find the range of values of $m$ for which the quadratic equation

$$(m - 1)x^2 + 2mx + (7m - 4) = 0 \quad \text{has no real roots.} \quad [5]$$

**9** The diagram below shows a sketch of the graph $y = f(x)$. The graph passes through the points $(-1, 0)$ and $(4, 0)$ and has a maximum point at $(2, 0)$.

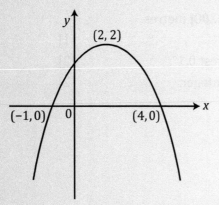

(a) Sketch the following graphs, using a separate set of axes for each graph. For each graph you should indicate the coordinates of the points of intersection with the $x$-axis and also indicate the coordinates of the stationary point.

(i) $y = f(x + 2)$ [3]

(ii) $y = -2f(x)$ [3]

(b) Hence write down one root of the equation

$$f(x + 2) = -2f(x) + 4$$ [2]

**10** The triangle ABC has AB = 16 cm, AC = 8 cm and angle ABC = 20°.

The triangle described above can be drawn in the following two ways, one with angle ACB as an obtuse angle and the other with angle ACB as an acute angle.

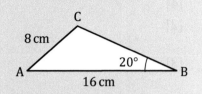 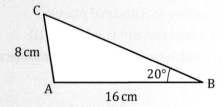

(a) Find the two possible sizes of angle ACB giving your answers to the nearest 0.1°. [3]

(b) Hence, find the two possible lengths of side BC giving your answer to three significant figures. [4]

**11** The number of radioactive nuclei, $N$, remaining after a time $t$ seconds may be modelled by $N = Ae^{-kt}$ where $A$ and $k$ are constants.

It is known that $N = 1000$ when $t = 4$ and that $N = 300$ when $t = 8$.

(a) Interpret the constant $A$ in the context of the question. [1]

(b) Find the value of $k$ giving your answer to 3 decimal places. [4]

(c) Find the number of radioactive nuclei remaining when $t = 10$. [2]

(d) Find the time, to the nearest second, when the number of radioactive nuclei remaining drops to 200. [3]

12 Given that $y = 20x^2 + 9x - 20$, find $\dfrac{dy}{dx}$ from first principles. [5]

**13** The diagram below shows a sketch of the curve $y = x^2 - 4x - 5$. The curve intersects the *x*-axis at points A and B. The tangent to the curve at D (4, −5) intersects the *x*-axis at point C.

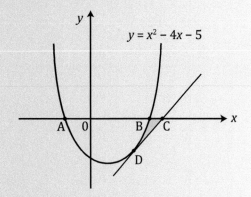

(a)  Find the coordinates of points A and B.                                    [2]

(b)  Find the equation of the tangent to the curve at D.              [4]

(c)  Find the area of the shaded region giving your answer as a mixed number.  [7]

**14** Circle C has equation $x^2 + y^2 - 8x + 10y + 28 = 0$.

Show that the line $2x + 3y = 6$ is a tangent to this circle and find the coordinates of the point of contact. [6]

⑮ An open glass tank is in the shape of a cuboid with length $2x$ cm, width $x$ cm and height $h$ cm.

The area of glass used to make the tank is $60\,000$ cm².

(a) Show that $h = \dfrac{30000 - x^2}{3x}$      [2]

(b) Show that $V = 20000x - \dfrac{2}{3}x^3$      [3]

(c) Find the value of $x$ that will give the tank its maximum volume and show that this value of $x$ will give the maximum volume.      [6]

**[Total marks for paper = 100]**

## Section A – Statistics

1. A teacher is concerned about the amount of time students in her class spend on social media per day. She decides to collect data by asking the students in her year 10 class about the time they spend. The results in hours per day for 25 students are shown below:

   0, 2, 1, 4, 2, 1, 5, 6, 3, 5, 1, 2, 0, 5, 4, 3, 2, 4, 6, 2, 1, 0, 1, 5, 2

   (a) The teacher decides to take an opportunity sample of the first 10 numbers in the list to calculate the mean.

      (i) Explain the meaning of the term 'opportunity sample'. [1]

      (ii) Work out the mean using the first 10 numbers. [1]

   (b) A systematic sample is to be taken of 5 data values.

      (i) Work out the sampling interval. [1]

      (ii) The teacher selects a random number in the sampling interval and it was 4. Using this value, write down the list of data values in the sample. [2]

      (iii) Using the list from (b)(ii), work out the value for the mean number of hours spent on social media. [1]

   (c) State, giving reasons, which of the two methods for working out the mean would be more likely to give a reliable result. [2]

**2** *A* and *B* are independent events such that

$P(A) = 0.2, \quad P(A \cup B) = 0.4$

(a) Determine the value of P(*B*). [4]

(b) Calculate the probability that exactly one of the events *A*, *B* occurs. [3]

**3** The heights of 42 students from college A were obtained and entered into a computer to produce the following box and whisker diagram:

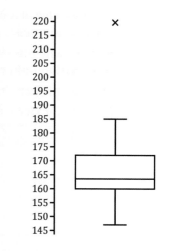

(a) There is an outlier on this diagram.

   (i) Explain what an outlier is and what has been done with the outlier when producing the diagram. [2]

   (ii) What will happen to the mean if the outlier is removed? [1]

   (iii) What will happen to the standard deviation if the outlier is removed? [1]

(b) The outlier is removed from the set of data and the remaining 41 heights were used to obtain the following summary statistics:

**College A Summary statistics**

| Height in cm | N | Mean | Standard deviation | Minimum | Lower quartile | Median | Upper quartile | Maximum |
|---|---|---|---|---|---|---|---|---|
| | 41 | 164.8 | 9.2 | 147 | 160 | 165 | 170.5 | 185 |

The heights of a set of 50 students from a different college, college B were obtained and the summary statistics for these are shown in the following table:

**College B Summary statistics**

| Height in cm | N | Mean | Standard deviation | Minimum | Lower quartile | Median | Upper quartile | Maximum |
|---|---|---|---|---|---|---|---|---|
| | 50 | 163.9 | 8.6 | 148 | 162 | 164 | 172 | 183 |

   (i) Work out the range and the IQR for each set of students. [2]

   (ii) Compare and contrast the distributions of heights of students for the two colleges A and B. [3]

**4** Customers arrive at a petrol station such that the number arriving in an interval of $t$ minutes is given by $0.5t$.

(a) Explain why this situation can be modelled using a Poisson distribution rather than a binomial distribution. [2]

(b) Find the probability that, between 9:30 am. and 10:00 am.,

    (i) exactly 18 customers arrive

    (ii) the number of customers arriving is greater than 20. [6]

**5** India is growing pumpkins for Halloween. From past experience she has found that there is a probability of 0.3 of a pumpkin having a radius of over 30 cm. India would like to see if the use of a new type of fertiliser specially developed for pumpkins will increase the number of large pumpkins.

India treats her pumpkins with the new fertiliser and takes a random sample of 40 pumpkins and she wants to conduct a test to see if the sizes of the pumpkins have increased.

(a) Write down a suitable hypothesis test India could use. [1]

(b) Find the critical region for the test at a 5% level of significance. [4]

(c) Write down the actual significance level of the test. [1]

**6** A random sample of 25 students was obtained in four different schools and their Maths and English marks in a GCSE mock examination were recorded and the following scatter graph plotted:

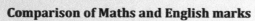

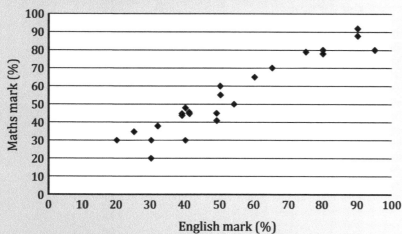

(a) (i) Comment on the correlation between 'Maths mark' and 'English mark'. [1]

    (ii) Interpret the correlation between the Maths and English marks. [1]

(b) The regression equation for this set of data is

    'Maths mark' = 7.1 + 0.9 × 'English mark'

    (i) Interpret the gradient of the regression equation for this model. [1]

    (ii) State, with a reason, whether the regression model would be useful to predict the Maths mark for a student who could not take the maths exam due to illness, but was present for the English exam. [2]

    (iii) State whether the relationship between the 'Maths mark' and 'English mark' is causal. Explain your answer. [1]

**[Total marks = 44]**

## Section B – Mechanics

**1** A hot air balloon rises vertically with a constant velocity of $2 \text{ m s}^{-1}$. When it is 50 m above the ground a sand bag is dropped from the balloon.

(a) Calculate the time the sand bag takes to hit the ground. [3]

(b) The loss of the sand bag causes the balloon to accelerate upwards with a constant acceleration of $2 \text{ m s}^{-2}$. Find the height of the balloon 4 seconds after releasing the sand bag. [3]

**2** (a) A particle accelerates for 10 seconds in a straight line so that $t$ seconds after starting its velocity, in m s$^{-1}$, is given by the formula

$$v = 3t^2 + 12$$

(i) Give the velocity of the particle when $t = 0$ s. [1]

(ii) Explain by drawing a sketch of the velocity–time graph, why the velocity is never negative. [2]

(b) A different particle has a velocity in m s$^{-1}$ given by $v = 13t + 8$.

(i) Explain why the acceleration of this particle is constant. [1]

(ii) Find the times when the two particles have the same velocity. [2]

**3** The diagram below shows two particles A and B, of mass *M* kg and 5 kg respectively where *M* > 5. Both particles are connected by a light inextensible string passing over a fixed smooth pulley. Initially, B is held at rest with the string just taut. It is then released.

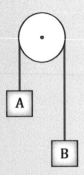

(a) Show that the acceleration of mass A is given by the equation
$$a = \frac{g(M-5)}{M+5}$$
[6]

(b) What modelling assumption did you make to arrive at your answer? [1]

**4** Two forces **P** and **Q** act on an object such that:

$$\mathbf{P} = 7\mathbf{i} + 14\mathbf{j} \qquad \mathbf{Q} = -2\mathbf{i} - 2\mathbf{j}$$

The object has a mass of 5 kg.

(a)  Show the two forces **P** and **Q** on a set of axes. [2]

(b)  Calculate the magnitude of the acceleration of the object. [3]

(c)  Calculate the direction of the acceleration giving your answer to one decimal place. [2]

**5** A person of mass 50 kg stands on the floor of a lift of mass 450 kg. The lift is raised and lowered by a metal cable.

The lift starts from rest and accelerates upwards at a constant rate until it reaches a speed of 4 m s⁻¹ after travelling a distance of 4 m from rest.

Find the tension in the lift cable. [4]

**[Total marks = 30]**

# Answers

## Unit 1  Pure Mathematics
## 1  Proof

**1** As there are a limited number of values to try, we use proof by exhaustion.

| $n$ | $n^2 + 2$ | Multiple of 4? |
|---|---|---|
| 1 | 3 | No |
| 2 | 6 | No |
| 3 | 11 | No |
| 4 | 18 | No |
| 5 | 27 | No |
| 6 | 38 | No |
| 7 | 51 | No |

For proof by exhaustion, it is a good idea to create a table like the one shown here.

As all the possible values have been exhausted and none of the answers are divisible by 4 exactly we have proved that $n^2 + 2$ is not a multiple of 4 for values of $n$ from 1 to 7 inclusive.

**2** Try $x = 2$ and $y = 1$ so $x^2 = 4$ and $y^2 = 1$.
Hence $x^2 > y^2$ is true.
Try $x = -1$ and $y = -2$ so so $x^2 = 1$ and $y^2 = 4$.
Hence $x^2 > y^2$ is false.
A counter-example has been found so the statement is false.

You have to find just one example of values where the statement is false to disprove the statement by counter-example.

**3** (a) Let $c = -1$     so $(2c + 1)^2 = 1$
Let $d = 0$     so $(2d + 1)^2 = 1$
Now $(2c + 1)^2 = (2d + 1)^2$ but $c \neq d$ so statement A is false by counter-example.

(b) Any real number has a unique, real cube root, so cube rooting both sides gives
$(2c + 1) = (2d + 1)$
Hence $c = d$ so statement B is true.

Try different values until you find an example where the statement is false.

Note that square rooting a positive real number results in two solutions. For example $\sqrt{25} = \pm 5$. However, a cube root has only one solution. For example $\sqrt[3]{27} = 3$.

**4** Let the first number be $2n$ (i.e. an even number).  The next number will be $2n + 1$ (i.e. an odd number).
Sum of the squares of these consecutive numbers
$= (2n)^2 + (2n + 1)^2 = 4n^2 + 4n^2 + 4n + 1 = 8n^2 + 4n + 1$
Now $8n^2 + 4n$ will always be even as it has 2 as a factor.  Adding 1 to it will make it odd.
Hence the sum of the squares of any two consecutive integers is always an odd number.

**5** If you list all the possible values of $n$ you have:   1, 2, 3, 4, 5, 6, 7, ...
All of these numbers are a multiple of 3 (i.e. 3 and 6), one less than a multiple of 3 (i.e. 2 and 5) or one more than a multiple of 3 (i.e. 4 and 7).
There are three situations to consider:
- If $n$ is a multiple of 3 then $n^3$ is a multiple of 27, and is therefore a multiple of 9.
  So if $n = 3x$, $n^3 = 27x^3$ so, as 9 is a factor of 27, $n^3$ is a multiple of 9.
- If $n$ is 1 more than a multiple of 3 then $n^3$ is 1 more than a multiple of 9.
  So if $n = 3x + 1$, $n^3 = (3x + 1)^3 = 27x^3 + 27x^2 + 9x + 1 = 9(3x^3 + 3x^2 + x) + 1$,
  which is 1 more than a multiple of 9.
- If $n$ is 1 less than a multiple of 3 then $n^3$ is 1 less than a multiple of 9.
  So if $n = 3x - 1$, $n^3 = (3x - 1)^3 = 27x^3 - 27x^2 + 9x - 1 = 9(3x^3 - 3x^2 + x) - 1$
  which is 1 less than a multiple of 9.

Hence the statement is true and it has been proved by proof by exhaustion.

# 2 Algebra and functions

**1** (a) $1 - 5x > -2x + 7$

$1 - 3x > 7$

$-3x > 6$

$x < -2$

Remember if you multiply or divide an inequality by a negative number, you must reverse the inequality sign.

(b) $\frac{x}{4} \leq 2(1 - x)$

$x \leq 8(1 - x)$

$x \leq 8 - 8x$

$9x \leq 8$

$x \leq \frac{8}{9}$

(c) $2x^2 + 5x - 12 \leq 0$

Put $(2x - 3)(x + 4) = 0$

$x = \frac{3}{2}$ or $-4$

The graph of $y = 2x^2 + 5x - 12$ will be U-shaped as the coefficient of $x^2$ is positive.

The section of the graph representing the inequality will be on or below the $y$-axis.

Hence solution is $-4 \leq x \leq \frac{3}{2}$

**2** (a) Let $f(x) = x^3 - 8x^2 - px + 84$

As $(x - 7)$ is a factor, $f(7) = 0$

$f(7) = (7)^3 - 8(7)^2 - p(7) + 84 = 0$

$343 - 392 - 7p + 84 = 0$

Giving $p = 5$

(b) $x^3 - 8x^2 - 5x + 84 = (x - 7)(ax^2 + bx + c)$

Equating coefficients of $x^3$ we obtain $a = 1$.

Equating coefficients independent of $x$ we obtain $84 = -7c$, hence $c = -12$

Equating coefficients of $x^2$ we obtain $-8 = b - 7a$ and as $a = 1$, $b = -1$

Substituting in these values we obtain

$(x - 7)(x^2 - x - 12)$

Factorising the quadratic factor gives

$(x - 7)(x - 4)(x + 3) = 0$

Solutions are $x = -3$ or $4$ or $7$

**3** (a) $1 - 2x < 4x + 7$

$1 - 6x < 7$

$-6x < 6$

$x > -1$

Dividing both sides by $-6$ and reversing the sign.

(b) $\frac{x}{2} \geq 2(1 - 3x)$

$x \geq 4(1 - 3x)$

$x \geq 4 - 12x$

$13x \geq 4$

$x \geq \frac{4}{13}$

**4** $f(x) = 2x^3 + 7x^2 - 7x - 12$

$f(1) = 2(1)^3 + 7(1)^2 - 7(1) - 12 = -10$

$f(-1) = 2(-1)^3 + 7(-1)^2 - 7(-1) - 12 = 0$ so $(x + 1)$ is a factor.

Here you find values for $x$ that will result in $f(x) = 0$.

The original function is a cubic function with three factors. One of the factors is $(x + 1)$ so the original function can be written like this:

$(x + 1)(ax^2 + bx + c) = 2x^3 + 7x^2 - 7x - 12$

Answers  Unit 1  Pure Mathematics

Equating the coefficients of $x^3$ gives, $a = 2$.
Equating the coefficients independent of $x$ gives, $c = -12$
Equating the coefficients of $x^2$ gives, $b + a = 7$, so $b = 5$.
Hence, $2x^3 + 7x^2 - 7x - 12 = (x + 1)(2x^2 + 5x - 12)$
Factorising the quadratic part into two factors gives:
$$(x + 1)(2x - 3)(x + 4)$$
Hence   $(x + 1)(2x - 3)(x + 4) = 0$

Solutions are $x = -1$, or $\frac{3}{2}$ or $-4$.

**5** (a) $(4 - 2\sqrt{5})(3 + 4\sqrt{5}) = 12 + 16\sqrt{5} - 6\sqrt{5} - 40$
$$= 10\sqrt{5} - 28$$

(b) $\sqrt{27} + \dfrac{81}{\sqrt{3}} = \sqrt{9 \times 3} + \dfrac{81\sqrt{3}}{\sqrt{3}\sqrt{3}} = 3\sqrt{3} + 27\sqrt{3}$
$$= 30\sqrt{3}$$

**6** $3x^2 + mx + 12 = 0$
For non-real roots   $b^2 - 4ac < 0$
So   $m^2 - 4(3)(12) < 0$
$m^2 - 144 < 0$
$(m + 12)(m - 12) < 0$

If a graph of $y = (m + 12)(m - 12)$ is plotted against $m$ on the $x$-axis the curve is U-shaped cutting the $x$-axis at $m = 12$ and $m = -12$.
The region needed is below the $x$-axis.
Hence the required range of $m$ is $-12 < m < 12$

> For no real roots, the discriminant is less than zero.

**7** (a) $\sqrt{27} + \sqrt{48} = \sqrt{9 \times 3} + \sqrt{16 \times 3} = 3\sqrt{3} + 4\sqrt{3} = 7\sqrt{3}$

(b) $\dfrac{20}{2 - \sqrt{2}} = \dfrac{20(2 + \sqrt{2})}{(2 - \sqrt{2})(2 + \sqrt{2})} = \dfrac{40 + 20\sqrt{2}}{4 - 2} = \dfrac{40 + 20\sqrt{2}}{2} = 20 + 10\sqrt{2}$

> Try to spot square numbers as factors.

**8** (a) $3x^2 - 12x + 10 = 3\left(x^2 - 4x + \dfrac{10}{3}\right)$
$$= 3\left((x - 2)^2 - 4 + \dfrac{10}{3}\right)$$
$$= 3\left((x - 2)^2 - \dfrac{12}{3} + \dfrac{10}{3}\right)$$
$$= 3\left((x - 2)^2 - \dfrac{2}{3}\right)$$
$$= 3(x - 2)^2 - 2$$

Hence minimum point is at $(2, -2)$

(b) The maximum value of $\dfrac{1}{3x^2 - 12x + 10}$ occurs when $3x^2 - 12x + 10$ has its smallest value.
The smallest value of $3x^2 - 12x + 10$ is the $y$-value of the minimum point which from part (a) is $-2$.

Hence the maximum value is $\dfrac{1}{-2} = -\dfrac{1}{2}$

> If you are asked to find the minimum or maximum value then we complete the square and find the turning point of the curve represented by the function.

> The graph with this equation will be U-shaped. The 3 only determines the shape of the U and does not alter where the minimum point is.

**9** (a) $\left(\dfrac{27}{9}\right)^0 = 1$

(b) $27^{\frac{2}{3}} = \sqrt[3]{27^2} = 3^2 = 9$

(c) $\left(\dfrac{27}{8}\right)^{-\frac{1}{3}} = \dfrac{1}{\sqrt[3]{\frac{27}{8}}} = \dfrac{1}{\frac{3}{2}} = \dfrac{2}{3}$

> The rules of indices must be remembered as you will be using them throughout the course in other topics.

104

**10** **(a)**

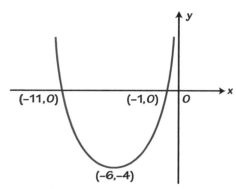

(–11,0)          (–1,0)     O

(–6,–4)

**(b)** Looking at the graph it has been reflected in the *x*-axis (so where it cuts the *x*-axis stays the same) and the *y*-value of the turning point has been halved.

Equation is $y = -\frac{1}{2}f(x)$

So $r = -\frac{1}{2}$

**11** The graph of $y = \frac{1}{x}$ is shown here.

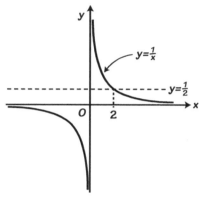

The line $y = \frac{1}{2}$ meets the graph of $y = \frac{1}{x}$ at $x = 2$.

As $\frac{1}{x} < \frac{1}{2}$ we need to find the values of *x* for which the graph of $y = \frac{1}{x}$ lies below the line $y = \frac{1}{2}$. From the graph we can see that this occurs for $x > 2$ or $x < 0$.

> You should be able to draw the graph of $y = \frac{1}{x}$ from your GCSE work.

**12** To find the *x*-coordinates of points Q and R:

$$y = -x^2 + 2x + 8$$

$$0 = x^2 - 2x - 8$$

$$0 = (x - 4)(x + 2)$$

Q is (4, 0) and R is (–2, 0)

To find the *y*-coordinate of P substitute $x = 0$, into the equation of the curve.

$$y = -x^2 + 2x + 8 = 0^2 + 2(0) + 8 = 8$$

Gradient of line L (i.e. PQ) $= -\frac{8}{4} = -2$

Equation of line L is $y = -2x + 8$

Looking at the shaded area we can see it lies below line L and below the curve and above the *x*-axis.

The three inequalities are:  $y \leq -2x + 8$,  $y \leq -x^2 + 2x + 8$,  $y \geq 0$.

> We need to find the equation of line L so we need to find the gradient, *m*, and the intercept, *c*, on the *y*-axis.

# 3 Coordinate geometry

**1** (a) $2y = 4x - 5$

$y = 2x - \dfrac{5}{2}$

Comparing this to $y = mx + c$ we have gradient, $m = 2$

(b) Gradient $= -\dfrac{1}{2}$

**2** (a) Gradient $= \dfrac{y_2 - y_1}{x_2 - x_1} = \dfrac{4 - 0}{6 - (-2)} = \dfrac{4}{8} = \dfrac{1}{2}$

(b) $\left(\dfrac{x_1 + x_2}{2}, \dfrac{y_1 + y_2}{2}\right) = \left(\dfrac{-2 + 6}{2}, \dfrac{0 + 4}{2}\right) = (2, 2)$

(c) (i) Gradient $= -2$ (i.e. invert $\frac{1}{2}$ and change the sign)

   (ii) $y - y_1 = m(x - x_1)$

   $y - 2 = -2(x - 2)$

   $y - 2 = -2x + 4$

   $y = -2x + 6$

**3** (a) $2x + 3y = 5$ so $y = -\dfrac{2}{3}x + \dfrac{5}{3}$, hence gradient $= -\dfrac{2}{3}$

(b) When $x = 0$, $y - 3 = 2$, giving $y = 5$. Hence S is $(0, 5)$

**4** (a) Gradient of AB $= -\dfrac{4}{5}$

(b) Substituting the coordinates of C into the equation of the line we obtain

   $4(-5) + 5(6) = 10$

   $-20 + 30 = 10$

   $10 = 10$

   As LHS = RHS, point C lies on the line.

(c) Gradient of line at right-angles to AB $= \dfrac{5}{4}$

   Equation of perpendicular line is $y - 6 = \dfrac{5}{4}(x + 5)$

   Hence $4y - 24 = 5x + 25$

   Equation is $4y = 5x + 49$

**5** Comparing the equation $x^2 + y^2 + 6x + 8y - 10 = 0$ with the equation $x^2 + y^2 + 2gx + 2fy + c = 0$ we can see $g = 3$, $f = 4$ and $c = -10$.

Centre A has co-ordinates $(-g, -f) = (-3, -4)$

Radius $= \sqrt{g^2 + f^2 - c} = \sqrt{(3)^2 + (4)^2 + 10} = \sqrt{35} = 5.92$.

**6** (a) (i) $2x + 5y = 40$

   Hence $5y = -2x + 40$

   $y = -\dfrac{2}{5}x + 8$

   Comparing this with the equation for a straight line, $y = mx + c$

   we obtain, gradient $= -\dfrac{2}{5}$

   (ii) The line parallel to $2x + 5y = 40$ will have gradient $-\dfrac{2}{5}$.

   Equation of line with gradient $-\dfrac{2}{5}$ and passing through the point P $(0, 6)$ is

   $y - y_1 = m(x - x_1)$

   $y - 6 = -\dfrac{2}{5}(x - 0)$

   $5y - 30 = -2x$

   Hence, equation of line is $5y + 2x = 30$

> Parallel lines have the same gradient.

> You could use the other method here where you let the equation of the parallel line be $2x + 5y = c$, and then substitute the coordinates of the points through which the line passes into the equation to find the value of $c$. Once this is done, this will be the required equation.

(b) The coordinates $(5, p)$ lie on the line, so these coordinates will satisfy the equation for the line.

So, $5y - 30 = -2x$

$5p - 30 = -2(5)$

$5p - 30 = -10$

$5p = 20$

$p = 4$

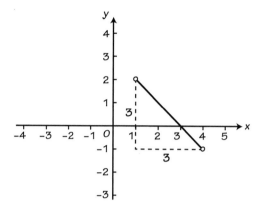

$x = 5$ and $y = p$ are substituted into the equation of the straight line and the resulting equation is solved to find the numerical value of $p$.

**7** First find the radius of the circle. You can either use the distance between two points formula or sketch the points on a set of axes and use Pythagoras' theorem.

Here we will use Pythagoras' theorem.

By Pythagoras' theorem, length of line $\sqrt{3^2 + 3^2} = \sqrt{18}$.

Equation of a circle having centre $(1, 2)$ and passing through the point $(4, -1)$ is $(x - 1)^2 + (y - 2)^2 = 18$.

**8** (a) $x^2 + y^2 - 4x + 8y + 4 = 0$

Completing the squares for $x$ and $y$ gives

$(x - 2)^2 + (y + 4)^2 - 4 - 16 + 4 = 0$

$(x - 2)^2 + (y + 4)^2 - 16 = 0$

$(x - 2)^2 + (y + 4)^2 = 16$

Centre of circle is at $(2, -4)$

(b) If the point P $(6, -4)$ lies on the circle, the coordinates will satisfy the equation of the circle.

Hence, $(x - 2)^2 + (y + 4)^2 = (6 - 2)^2 + (-4 + 4)^2 = 4^2 + 0 = 16$

This is the same as the RHS of the equation, so the point lies on the circle.

**9** (a) If AP is the diameter of the circle, then AB is a radius of the circle.

A graph is drawn to show the points.

Using Pythagoras' theorem to find AB we have

$AB^2 = 2^2 + 1^2$

$AB^2 = 4 + 1$

$AB = \sqrt{5}$

Radius of circle = $\sqrt{5}$ and centre is at B$(1, 2)$

Equation of the circle is $(x - 1)^2 + (y - 2)^2 = (\sqrt{5})^2$

$x^2 - 2x + 1 + y^2 - 4y + 4 = 5$

$x^2 + y^2 - 2x - 4y = 0$

Hence, $a = -2$, $b = -4$, $c = 0$

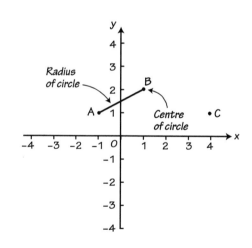

(b) To find the coordinates of P, the other end of the diameter of the circle we can use the fact that to go from A to B you go two units to the right and one unit vertically up. Applying this from B$(1, 2)$ gives the coordinates of P as $(3, 3)$.

Gradient of BP $= \dfrac{3 - 2}{3 - 1} = \dfrac{1}{2}$  (Note that BP is a radius of the circle)

Gradient of CP $= \dfrac{3 - 1}{3 - 4} = -2$

Product of the gradients $= \left(\dfrac{1}{2}\right)(-2) = -1$

Hence radius at P and line at P are perpendicular so CP is a tangent to the circle.

**10** (a) $AB^2 = AP^2 + BP^2$

$= (3\sqrt{5})^2 + (4\sqrt{5})^2$

$= 45 + 80$

$= 125$

$AB = \sqrt{125} = 5\sqrt{5}$

(b)   Equation of the circle is

$$(x - 3)^2 + (y + 2)^2 = \left(5\sqrt{5}\right)^2$$
$$(x - 3)^2 + (y + 2)^2 = 125$$
$$x^2 + y^2 - 6x + 4y - 112 = 0$$

**11**   To find the coordinates where the line meets the curve we can equate the $y$-values.

$$m(x - 1) = x^2 + 3$$
$$mx - m = x^2 + 3$$
$$x^2 - mx + 3 + m = 0$$

For the line to be a tangent this must have equal roots, as the line and the curve must have exactly one point of intersection.

For equal roots, the discriminant = 0

$$b^2 - 4ac = 0$$
$$(-m)^2 - 4(3 + m) = 0$$
$$m^2 - 12 - 4m = 0$$
$$m^2 - 4m - 12 = 0$$
$$(m - 6)(m + 2) = 0$$

So $m = 6$ or $-2$

**12**   (a)   Mid-point of the diameter PQ is $\left(\dfrac{1 + 3}{2}, \dfrac{3 + (-1)}{2}\right) = (2, 1)$ and

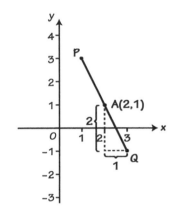

these are also the coordinates of point A, the centre of the circle.

By Pythagoras' theorem          $AQ^2 = 2^2 + 1^2$
$$AQ = \sqrt{5}$$

AQ is the radius of C, so radius = $\sqrt{5}$

Equation of circle C is   $(x - 2)^2 + (y - 1)^2 = \left(\sqrt{5}\right)^2$
$$x^2 - 4x + 4 + y^2 - 2y + 1 = 5$$
$$x^2 + y^2 - 4x - 2y = 0$$

Hence $a = -4$, $b = -2$ and $c = 0$

(b)

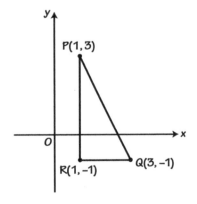

As points P and R have the same $x$-coordinates PR is parallel to the $y$-axis and as Q and R have the same $y$-coordinates, QR is parallel to the $x$-axis. This means angle PRQ = 90°.

Now PR = 4 and QR = 2, hence angle PQR = $\tan^{-1}\left(\dfrac{4}{2}\right)$ = 63.4° (1 d.p.)

# 4 Sequences and series

**1** First obtain the formula for the binomial expansion from the formula booklet.

$$(a+b)^n = a^n + \binom{n}{1}a^{n-1}b + \binom{n}{2}a^{n-2}b^2 + \ldots$$

Here $a = 3$, $b = 2x$ and $n = 3$.

Putting these values into the formula gives:

$$(3+2x)^3 = 3^3 + \binom{3}{1}3^2(2x) + \binom{3}{2}3^1(2x)^2 + \binom{3}{3}3^0(2x)^3$$

As $n = 3$ here we look for the line in Pascal's triangle which starts at 1 and then 3, etc.

You can see that the numbers in this line are:   1   3   3   1

These represent the numbers that can represent the factorials.

So, for example $\binom{3}{1} = 3$ and $\binom{3}{3} = 1$.

Hence, we can write the expansion like this:

$$(3+2x)^3 = (1)3^3 + (3)3^2(2x) + (3)3^1(2x)^2 + (1)3^0(2x)^3$$

Hence $(3+2x)^3 = 27 + 54x + 36x^2 + 8x^3$

> Remember that $3^0 = 1$

**2** The formula is as follows:

$$(a+b)^n = a^n + \binom{n}{1}a^{n-1}b + \binom{n}{2}a^{n-2}b^2 + \ldots + \binom{n}{r}a^{n-r}b^r + \ldots + b^n$$

Here $n = 6$, $a = x$ and $b = \dfrac{3}{x}$.

Substituting in the values for $a$, $b$ and $n$ we obtain

$$\left(x+\frac{3}{x}\right)^6 = x^6 + \binom{6}{1}x^5\left(\frac{3}{x}\right) + \binom{6}{2}x^4\left(\frac{3}{x}\right)^2 + \binom{6}{3}x^3\left(\frac{3}{x}\right)^3 + \ldots$$

Looking at the above it can be seen that the term in $x^2$ is the third term in the expansion.

Term in $x^2 = \binom{6}{2}x^4\left(\dfrac{3}{x}\right)^2$

To find the coefficients we will expand Pascal's triangle.

> Rather than use Pascal's triangle, you could use a calculator to find the value of $\binom{6}{2}$

```
            1
          1   1
        1   2   1
      1   3   3   1
    1   4   6   4   1
  1   5  10  10   5   1
1   6  15  20  15   6   1
```

The last line of Pascal's triangle shows the line we need as we need the second number in the line to be a 6 which is the power to which the bracket is to be raised.

As $\binom{6}{2} = 15$, we have the term in $x^2 = 15x^4\left(\dfrac{3}{x}\right)^2 = 135x^2$

**3** (a) The formula for the expansion of $(1+x)^n$ is obtained from the formula booklet.

$$(1+x)^n = 1 + nx + \frac{n(n-1)x^2}{2!} + \frac{n(n-1)(n-2)x^3}{3!} + \ldots$$

Putting $n = 7$ into this formula gives:

$$(1+x)^7 = 1 + 7x + \frac{7(6)x^2}{2!} + \frac{7(6)(5)x^3}{3!} + \ldots$$

Note that using the first three terms only provides an approximate value.

Hence,    $(1+x)^7 \approx 1 + 7x + \dfrac{7(6)x^2}{2!} + \dfrac{7(6)(5)x^3}{3!}$

$$\approx 1 + 7x + 21x^2 + 35x^3$$

(b)   $(1 + x)^7 \approx 1 + 7x + 21x^2 + 35x^3$

Let $x = 0.1$

$$(1 + 0.1)^7 \approx 1 + 7(0.1) + 21(0.1)^2 + 35(0.1)^3 \approx 1.945$$

$x = 0.1$ is substituted for $x$ into the expansion.

(c)   You could use the expansion and substitute $x = -0.01$ into it.
Note that $(1 - 0.01)^7 = (0.99)^7$.

④   (a)   $(a + b)^6 = a^6 + 6a^5b + 15a^4b^2 + 20a^3b^3 + 15a^2b^4 + 6ab^5 + b^6$
As $a = 1$ and $b = x$, we have:
$$(1 + x)^6 = 1^6 + 6(1^5)(x) + 15(1^4)(x^2) + 20(1^3)(x^3) + 15(1^2)(x^4) + 6(1)(x^5) + x^6$$
$$= 1 + 6x + 15x^2 + 20x^3 + 15x^4 + 6x^5 + x^6$$

Note we have used Pascal's triangle here to determine the coefficients in the expansion, which are 1 6 15 20 15 6 1. You could have alternatively used the formula to find these.

(b)   $(1.02)^6 = (1 + 0.02)^6$
Hence, we substitute $x = 0.02$ into the expansion from part (a).
$$(1 + 0.02)^6 = 1 + 6(0.02) + 15(0.02)^2 + 20(0.02)^3 + 15(0.02)^4 + 6(0.02)^5 + (0.02)^6$$
$$= 1.1262 \text{ (four decimal places)}$$

# 5 Trigonometry

①   $(2\cos\theta - 1)(\cos\theta + 1) = 0$
$$2\cos\theta - 1 = 0$$
$$2\cos\theta = 1$$
$$\cos\theta = \frac{1}{2}$$

We take each bracket and put the contents equal to zero.

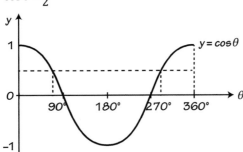

$\theta = 60°, 300°$
$$\cos\theta + 1 = 0$$
$$\cos\theta = -1$$
$$\theta = 180°$$
Hence, the values of $\theta$ in the range are $60°, 180°, 300°$

You could have used the CAST method to work out the angles.

②   $3\cos^2\theta - \cos\theta - 2 = 0$
Factorising, we obtain $(3\cos\theta + 2)(\cos\theta - 1) = 0$
Hence $3\cos\theta + 2 = 0$   or   $\cos\theta - 1 = 0$

$\cos\theta = -\frac{2}{3}$   or   $\cos\theta = 1$

It is important to spot that this is a quadratic equation in $\cos\theta$. To solve this equation we factorise it and then equate each of the brackets to zero. We then solve each of the resulting equations.

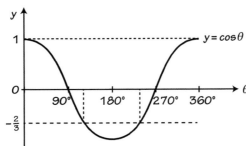

$\theta = 131.8°$ or $228.2°$   or   $\theta = 0°$ or $360°$
Hence values of $\theta$ are $0°, 131.8°, 228.2°, 360°$

Draw a graph of $y = \cos\theta$ to find the angles or alternatively use the CAST method.

Always check the range for the allowable values of $\theta$.

**3** (a) $3 \sin \theta = 1$, so $\sin \theta = \frac{1}{3}$ and $\theta = \sin^{-1}\frac{1}{3}$

$\theta = 19.5°, 160.5°$

(b) $\tan \theta = \frac{\sqrt{3}}{2}$ so $\theta = \tan^{-1}\left(\frac{\sqrt{3}}{2}\right)$

$\theta = 40.9°, 220.9°$

> Use either the symmetry of the graph or the CAST method to find the angles. Make sure you only include the angles in the range specified in the question.

(c) $3 \cos 2\theta = -1$

$\cos 2\theta = -\frac{1}{3}$

$2\theta = \cos^{-1}\left(-\frac{1}{3}\right)$

$2\theta = 109.5°, \ 250.5°, \ 469.5°, \ 610.5°$

$\theta = 54.75°, \ 125.25°, \ 234.75°, \ 305.25°$

(d)      $2 \cos^2 \theta + \sin \theta - 1 = 0$

$2(1 - \sin^2 \theta) + \sin \theta - 1 = 0$

$2 - 2 \sin^2 \theta + \sin \theta - 1 = 0$

$2 \sin^2 \theta - \sin \theta - 1 = 0$

$(2 \sin \theta + 1)(\sin \theta - 1) = 0$

$\sin \theta = -\frac{1}{2}$ or $\sin \theta = 1$

$\theta = 90°, \ 210°, \ 330°$

> Spot that this is a quadratic equation in $\sin \theta$.

**4** (a) $6 \sin^2 \theta + 1 = 2(\cos^2 \theta - \sin \theta)$

$6 \sin^2 \theta + 1 = 2(1 - \sin^2 \theta - \sin \theta)$

$6 \sin^2 \theta + 1 = 2 - 2 \sin^2 \theta - 2 \sin \theta$

$8 \sin^2 \theta + 2 \sin \theta - 1 = 0$

$(4 \sin \theta - 1)(2 \sin \theta + 1) = 0$

$\sin \theta = \frac{1}{4}$ or $\sin \theta = -\frac{1}{2}$

$\theta = \sin^{-1}\left(\frac{1}{4}\right)$ giving $\theta = 14.48°, 165.2°$ or

$\theta = \sin^{-1}\left(-\frac{1}{2}\right)$ giving $\theta = 210°, 330°$

Hence $\theta = 14.48°, 165.2°, 210°, 330°$

> Need to form a quadratic equation in $\sin \theta$ so $\cos^2 \theta$ is turned into $\sin^2 \theta$ using $\cos^2 \theta = 1 - \sin^2 \theta$.

> Remember to only give values in your final answer in the range specified in the question. Marks may be subtracted for values outside the range.

(b) $\tan (3x - 57°) = -0·81$

$3x - 57 = \tan^{-1}(-0·81)$

$3x - 57 = -39°, 141°, 321°, 501°$

$3x = 18°, 198°, 378°, 558°$

$x = 6°, 66°, 126°, 186°$

Hence     $x = 6°, 66°, 126°$

> Note that 186° is out of range so it is ignored.

(c) The minimum value of a sin or cos function is $-1$.

Hence $\sin \phi \geq -1$ and $\cos \phi \geq -1$, so $2 \sin \phi + 4 \cos \phi > -7$

This means there are no values satisfying $2 \sin \phi + 4 \cos \phi = -7$

**5** (a) Using the cosine rule we obtain:

$(x + 5)^2 = x^2 + 7^2 - 2 \times x \times 7 \cos \text{BAC}$

$x^2 + 10x + 25 = x^2 + 49 - 14x \cos \text{BAC}$

$x^2 + 10x + 25 = x^2 + 49 - 14x\left(-\frac{3}{5}\right)$

$x^2 + 10x + 25 = x^2 + 49 + 8.4x$

$1.6x = 24$

$x = 15 \text{ cm}$

(b)   Area of triangle $= \frac{1}{2}ab\sin C$

$$= \frac{1}{2} \times 15 \times 7 \times \sin \text{BAC}$$

Now $\sin \text{BAC} = \frac{4}{5}$

Area of triangle $= \frac{1}{2} \times 15 \times 7 \times \frac{4}{5}$

$$= 42\,\text{cm}^2$$

(c)   $42 = \frac{1}{2} \times 20 \times \text{AD}$

AD $= 4.2$ cm

**6**   (a)   Using the cosine rule
$$\text{AC}^2 = 8^2 + 15^2 - 2 \times 8 \times 15\cos 60°$$

$$\text{AC}^2 = 64 + 225 - 240 \times \frac{1}{2}$$

$$\text{AC}^2 = 169$$

AC $= 13$ cm

(b)   Using the sine rule   $\dfrac{b}{\sin B} = \dfrac{c}{\sin C}$

$$\frac{13}{\sin 60°} = \frac{8}{\sin \theta}$$

$$\sin \theta = \frac{8 \times \sin 60°}{13}$$

$\theta = 32.2°$ (nearest 0.1°)

> Here we know the sides and the included angle and want to find the other side. In this situation, the cosine rule is used.

# 6 Exponentials and logarithms

**1**   $\log_3 \dfrac{1}{2^3} + \log_3 27 + 3 = \log_3 2^{-3} + \log_3 3^3 + 3\log_3 3$

$$= -3\log_3 2 + 3\log_3 3 + 3\log_3 3$$

$$= 6 - 3\log_3 2$$

> Now $\log_3 3 = 1$

**2**   Now $1 = \log_2 2$

$$\log_2 36 - \log_2 15 + \log_2 100 + 1 = \log_2\left(\frac{36 \times 100 \times 2}{15}\right)$$

$$= \log_2 480$$

**3**   $3\log_{10} 4 - \dfrac{1}{2}\log_{10} 64 + 1 = \log_{10} 64 - \log_{10} 8 + \log_{10} 10$

$$= \log_{10}\frac{64 \times 10}{8}$$

$$= \log_{10} 80$$

**4**   Let $p = \log_3 a$   so $a = 3^p$
Let $q = \log_a 15$   so $a^q = 15$
Now $a^q = (3^p)^q = 3^{pq}$
But $a^q = 15$
Hence $3^{pq} = 15$
Taking logs to base 3 of both sides we obtain
$$\log_3 3^{pq} = \log_3 15$$
$$pq\log_3 3 = \log_3 15$$
Now $\log_3 3 = 1$
Hence $pq = \log_3 15$
Hence $\log_3 a \times \log_a 15 = \log_3 15$

**5** $2^{3-2x} = 5$

Taking logs to base 10 of both sides

$$\log_{10} 2^{3-2x} = \log_{10} 5$$

$$(3 - 2x)\log_{10} 2 = \log_{10} 5$$

$$(3 - 2x) = \frac{\log_{10} 5}{\log_{10} 2}$$

$$(3 - 2x) = 2.322$$

$$0.678 = 2x$$

$$x = 0.34 \ (2 \ \text{d.p.})$$

The rule $\log_{10} x^k = k\log_{10} x$ is used here

Be careful here; $\dfrac{\log_{10} 5}{\log_{10} 2}$

is not the same as $\log_{10} \dfrac{5}{2}$

**6** $9^x - 5(3^x) + 6 = 0$

Let $y = 3^x$     so $y^2 = 3^{2x}$

Hence, we can write          $9^x - 5(3^x) + 6 = 0$

as          $y^2 - 5y + 6 = 0$

$$(y - 3)(y - 2) = 0$$

Hence $y = 3$ or $2$

When $y = 3$, $3 = 3^x$, so $x = 1$

When $y = 2$, $2 = 3^x$

Taking logs of both sides

$$\log_{10} 2 = \log_{10} 3^x$$

$$\log_{10} 2 = x\log_{10} 3$$

$$x = \frac{\log_{10} 2}{\log_{10} 3} = 0.63$$

Remember to give the answer correct to the number of decimal places or significant figures.

If you look at this equation carefully you should spot that it is possible to turn it into a quadratic equation.

# 7 Differentiation

**1** Increasing $x$ by a small amount $\delta x$ will result in $y$ increasing by a small amount $\delta y$.

Substituting $x + \delta x$ and $y + \delta y$ into the equation we have:

$$y + \delta y = (x + \delta x)^3 - 5(x + \delta x)$$

Now $(x + \delta x)^3 = (x + \delta x)(x^2 + 2x\delta x + (\delta x)^2)$

$$= x^3 + 2x^2\delta x + x(\delta x)^2 + x^2\delta x + 2x(\delta x)^2 + (\delta x)^3$$

$$= x^3 + 3x^2\delta x + 3x(\delta x)^2 + (\delta x)^3$$

Hence   $y + \delta y = x^3 + 3x^2\delta x + 3x(\delta x)^2 + (\delta x)^3 - 5(x + \delta x)$

$$= x^3 + 3x^2\delta x + 3x(\delta x)^2 + (\delta x)^3 - 5x - 5\delta x$$

But $y = x^3 - 5x$

Subtracting these equations gives

$$\delta y = 3x^2\delta x + 3x(\delta x)^2 + (\delta x)^3 - 5\delta x$$

Dividing both sides by $\delta x$

$$\frac{\delta y}{\delta x} = 3x^2 + 3x\delta x + (\delta x)^2 - 5$$

Letting $\delta x \rightarrow 0$

$$\frac{dy}{dx} = \lim_{\delta x \to 0} \frac{\delta y}{\delta x} = 3x^2 - 5$$

**2** $y = \sqrt[3]{x^2} + \dfrac{64}{x}$

$y = x^{\frac{2}{3}} + 64x^{-1}$

$$\frac{dy}{dx} = \frac{2}{3}x^{-\frac{1}{3}} - 64x^{-2} = \frac{2}{3\sqrt[3]{x}} - \frac{64}{x^2}$$

When $x = 8$,   $\dfrac{dy}{dx} = \dfrac{2}{3\sqrt[3]{8}} - \dfrac{64}{8^2} = \dfrac{1}{3} - 1 = -\dfrac{2}{3}$

**3** (a) Width = $x$ so length = $25 - x$
$$\text{Area} = x(25 - x)$$
$$= 25x - x^2$$

(b) (i) $A = 25x - x^2$
$$\frac{dA}{dx} = 25 - 2x$$

The maximum value of $A$ occurs when $\frac{dA}{dx} = 0$
$$25 - 2x = 0$$

Hence $x = 12.5$ m

The width is 12.5 m and the length is $25 - 12.5 = 12.5$ m

(ii) Area = $12.5 \times 12.5 = 156.25$ m²

**4** First find the gradient
$$f'(x) = \frac{3x^2}{3} - 2x - 8$$
$$= x^2 - 2x - 8$$

Now we need to find the values of $x$ which will make this expression less than zero.
First find the roots of the equation.
$$x^2 - 2x - 8 = 0$$
$$(x - 4)(x + 2) = 0$$

From the graph we want those sections of the curve that are below (but not on) the $x$-axis.
Hence range of values for which $f(x)$ is a decreasing function are $-2 < x < 4$.

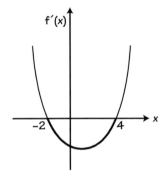

# 8 Integration

**1** $\int(5x^4 + 4x^3 - 2x^2 + x - 1)dx$
$$= \frac{5x^5}{5} + \frac{4x^4}{4} - \frac{2x^3}{3} + \frac{x^2}{2} - x + c$$
$$= x^5 + x^4 - \frac{2x^3}{3} + \frac{x^2}{2} - x + c$$

**2** $(x - 1)(x + 8) = x^2 + 8x - x - 8$
$$= x^2 + 7x - 8$$
$$\int(x - 1)(x + 8)dx = \int(x^2 + 7x - 8)\,dx$$
$$= \frac{x^3}{3} + \frac{7x^2}{2} - 8x + c$$

**3** $y = \int(3x^2 - 10x + 4)dx$
$$= \frac{3x^3}{3} - \frac{10x^2}{2} + 4x + c$$
$$= x^3 - 5x^2 + 4x + c$$

Now when $x = 2$, $y = 4$, so substituting these values into the above equation we obtain
$$4 = (2)^3 - 5(2)^2 + 4(2) + c$$

Solving gives $c = 8$

Hence $y = x^3 - 5x^2 + 4x + 8$

**4** $\int\left(\frac{x^2}{5} + \frac{x}{2}\right)dx = \frac{x^3}{5 \times 3} - \frac{x^2}{2 \times 2} + c$
$$= \frac{x^3}{15} - \frac{x^2}{4} + c$$

> Notice the way when you increase the index by one and then divide by the new index, if there is a number already in the denominator the new number is multiplied by it.

**5** $\displaystyle\int_0^1 \frac{2}{3}(5x-6)\,dx = \frac{2}{3}\int_0^1(5x-6)\,dx$

$\displaystyle\frac{2}{3}\int_0^1(5x-6)\,dx = \frac{2}{3}\left[\frac{5x^2}{2}-6x\right]_0^1$

$\displaystyle\qquad\qquad\qquad = \frac{2}{3}\left[\left(\frac{5(1)^2}{2}-6(1)\right)-(0)\right]$

$\displaystyle\qquad\qquad\qquad = \frac{2}{3}(-3.5)$

$\displaystyle\qquad\qquad\qquad = -2.33$

> Note that you can remove the fraction outside the integral sign which will make the integration easier.

**6** (a) To find where the curve cuts the $x$-axis we substitute $y = 0$ into the equation of the curve.
Hence $x^2 - 4 = 0$
$(x-2)(x+2) = 0$
Solving, gives $x = 2$ or $-2$.
Note as the curve has a positive coefficient of $x^2$ the curve will be $\cup$-shaped.
Sketching the curve we obtain the graph shown.

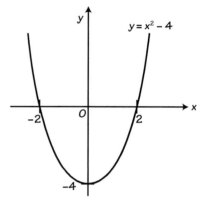

(b) $\displaystyle\int_2^3(x^2-4)\,dx = \left[\frac{x^3}{3}-4x\right]_2^3$

$\displaystyle\qquad\qquad\qquad = \left[\left(\frac{3^3}{3}-4(3)\right)-\left(\frac{2^3}{3}-4(2)\right)\right]$

$\displaystyle\qquad\qquad\qquad = (9-12)-\left(\frac{8}{3}-8\right)$

$\displaystyle\qquad\qquad\qquad = \frac{7}{3}$

$\displaystyle\int_0^2(x^2-4)\,dx = \left[\frac{x^3}{3}-4x\right]_0^2$

$\displaystyle\qquad\qquad\qquad = \left[\left(\frac{2^3}{3}-4(2)\right)-(0)\right]$

$\displaystyle\qquad\qquad\qquad = -\frac{16}{3}$

(c) The positive integral represents the area above the $x$-axis and the negative area represents the area below the $x$-axis.

**7** (a) $\displaystyle\int\left(\frac{3}{\sqrt[4]{x}}-9x^{\frac{5}{2}}\right)dx = \int\left(3x^{-\frac{1}{4}}-9x^{\frac{5}{2}}\right)dx$

$\displaystyle\qquad\qquad\qquad = \frac{3x^{\frac{3}{4}}}{\frac{3}{4}}-\frac{9x^{\frac{7}{2}}}{\frac{7}{2}}+c$

$\displaystyle\qquad\qquad\qquad = 4x^{\frac{3}{4}}-\frac{18x^{\frac{7}{2}}}{7}+c$

> Need to change the $\frac{3}{\sqrt[4]{x}}$ term into index form ready for integrating.

(b) Area $= \displaystyle\int_1^4(2x^2+6x^{-2})\,dx$

$\displaystyle\qquad = \left[\frac{2x^3}{3}+\frac{6x^{-1}}{-1}\right]_1^4$

$\displaystyle\qquad = \left[\frac{2x^3}{3}-\frac{6}{x}\right]_1^4$

$\displaystyle\qquad = \left[\left(\frac{2\times 4^3}{3}-\frac{6}{4}\right)-\left(\frac{2\times 1^3}{3}-\frac{6}{1}\right)\right]$

$\displaystyle\qquad = 46.5$ square units

**8** (a) $\frac{dy}{dx} = x^2 + 2x - 8$

So, $y = \int (x^2 + 2x - 8)dx$

Integrating both sides with respect to $x$ gives

$$y = \frac{x^3}{3} + \frac{2x^2}{2} - 8x + c$$

$$y = \frac{x^3}{3} + x^2 - 8x + c$$

Now as the point $(3, 0)$ lies on the curve the coordinates will satisfy the equation of the curve. So,

$$0 = \frac{(3)^3}{3} + (3)^2 - 8(3) + c$$

Solving gives $c = 6$.

Hence $y = \frac{x^3}{3} + x^2 - 8x + 6$

> Remember when you integrate the derivative, you obtain the original equation provided you are able to find the value of $c$, the constant of integration.

(b) At the stationary points, $\frac{dy}{dx} = 0$, so $x^2 + 2x - 8 = 0$.

Factorising the quadratic equation gives $(x - 2)(x + 4) = 0$.

Hence, $x = 2$ or $-4$

When $x = 2$,    $y = \frac{(2)^3}{3} + (2)^2 - 8(2) + 6 = -3\frac{1}{3}$

When $x = -4$,    $y = \frac{(-4)^3}{3} + (-4)^2 - 8(-4) + 6 = 32\frac{2}{3}$

> If the question wanted just the $x$-coordinate it would have specified it. If you are asked for the coordinates of the stationary points, you must give both the $x$- and $y$-coordinates.

(c) Finding the intercept on the $y$-axis by substituting $x = 0$ into the equation of the curve we have

$y = \frac{(0)^3}{3} + (0)^2 - 8(0) + 6 = 6$.

Adding the points to the sketch we have the graph shown.

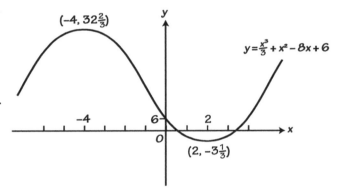

**9** (a) Need to first find the gradient of the tangent at B.

$$\frac{dy}{dx} = 3 - 2x$$

When $x = 2$, $\frac{dy}{dx} = 3 - 2(2) = -1$

Equation of tangent at B is $y - 2 = -1(x - 2)$

$$y = -x + 4$$

(b) To find the coordinate of C substitute $y = 0$ in the equation of the tangent.

$0 = -x + 4$, so $x = 4$ and C is the point $(4, 0)$

To find the coordinates of point A we substitute the equation of the $x$-axis (i.e. $y = 0$) into the equation of the curve.

So,    $3x - x^2 = 0$

$x(3 - x) = 0$

$x = 0$ or $3$ so A must be the point with $x$-coordinate $= 3$

Hence A has coordinates $(3, 0)$

Area of triangle BCD $= \frac{1}{2} \times$ CD $\times$ BD $= \frac{1}{2} \times 2 \times 2 = 2$ square units

Area under the curve between A and D $= \int_2^3 y\,dx$

$= \int_2^3 (3x - x^2)dx$

$= \left[\frac{3x^2}{2} - \frac{x^3}{3}\right]_2^3$

$= \left[\left(\frac{3(3)^2}{2} - \frac{(3)^3}{3}\right) - \left(\frac{3(2)^2}{2} - \frac{(2)^3}{3}\right)\right] = \left[\left(\frac{27}{2} - 9\right) - \left(6 - \frac{8}{3}\right)\right] = \frac{7}{6}$ square units

Shaded area    $= 2 - \frac{7}{6} = \frac{5}{6}$ square units

# 9 Vectors

**1** (a) $2\mathbf{a} - \mathbf{b} = 2(4\mathbf{i} - 3\mathbf{j}) - (-2\mathbf{i} + 5\mathbf{j})$
$$= 8\mathbf{i} - 6\mathbf{j} + 2\mathbf{i} - 5\mathbf{j}$$
$$= 10\mathbf{i} - 11\mathbf{j}$$

(b)

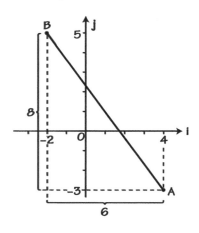

By Pythagoras' theorem
$AB^2 = 8^2 + 6^2$
$AB = 10$

> An alternative method would be to express A and B as (4, −3) and (−2, 5) respectively and use the formula to find the distance between two points.

**2**

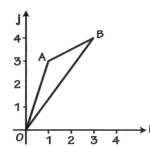

> It is always worth spending the time drawing a diagram when answering vector questions.

(a) $\mathbf{AB} = \mathbf{AO} + \mathbf{OB}$
$$= (-\mathbf{i} - 3\mathbf{j}) + (3\mathbf{i} + 4\mathbf{j})$$
$$= 2\mathbf{i} + \mathbf{j}$$

(b) $|\mathbf{AB}| = \sqrt{2^2 + 1} = \sqrt{5}$

> Remember that vectors can be written in bold rather than writing them with an arrow over them. Be prepared for either being used in the exam.

**3**

> Don't work this out using a calculator as the exact value is needed.

(a) Gradient of AB $= \dfrac{3-2}{2.5-1} = \dfrac{1}{1.5} = \dfrac{2}{3}$

Gradient of BC $= \dfrac{3-0}{2.5-4.5} = -\dfrac{3}{2}$

Product of gradients $= \dfrac{2}{3} \times -\dfrac{3}{2} = -1$ so AB and BC are perpendicular

(b) $\mathbf{AC} = \mathbf{AB} + \mathbf{BC}$
$\mathbf{AC} = (1.5\mathbf{i} + \mathbf{j}) + (2\mathbf{i} - 3\mathbf{j})$
$$= 3.5\mathbf{i} - 2\mathbf{j}$$

> An alternative method to this would be to use the position vectors to work out **AB** and **AC**, e.g. **AB** = **AO** + **OB**

**4**   (a)   **AB = AO + OB**
    = (2**i** + **j**) + (4**i** + **j**)
    = 6**i** + 2**j**

  (b)   **DC** = 6**i** + 2**j**
    As **AB** and **DC** are identical vectors they are parallel and the same length.

> To go from D to C you go 6 units in the positive **i** direction and 2 units in the positive **j** direction.

**5**   (a)   (i)   2**a** – 3**b** = 2(9**i** – 2**j**) – 3(–3**i** + 3**j**)
        = 18**i** – 4**j** + 9**i** – 9**j**
        = 27**i** – 13**j**

  (ii)   The coordinates of P and Q are (9, –2) and (–3, 3) respectively.

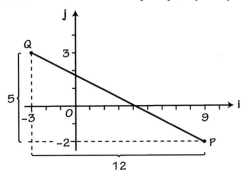

> You could alternatively have used the formula for the distance between two points:
> $$d = \sqrt{(x_2 - x_1)^2 + (y_2 - y_1)^2}$$
> $$PQ = \sqrt{(-3 - 9)^2 + (3 - (-2))^2}$$
> $$PQ = 13$$

    By Pythagoras' theorem     $PQ^2 = 12^2 + 5^2$
                $PQ = 13$

  (b)   (i)

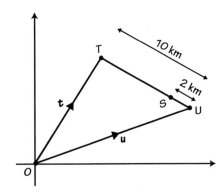

Position vector of S = $\overrightarrow{OS}$
$\overrightarrow{OS} = \overrightarrow{OU} + \overrightarrow{US}$
Now $\overrightarrow{US} = \frac{1}{5}\overrightarrow{UT}$ and $\overrightarrow{UT} = \overrightarrow{UO} + \overrightarrow{OT} = \mathbf{t} - \mathbf{u}$
Hence, $\overrightarrow{OS} = \overrightarrow{OU} + \overrightarrow{US} = \mathbf{u} + \frac{1}{5}(\mathbf{t} - \mathbf{u}) = \frac{4}{5}\mathbf{u} + \frac{1}{5}\mathbf{t}$
Position vector of S = $\frac{4}{5}\mathbf{u} + \frac{1}{5}\mathbf{t}$

  (ii)   Looking at the position vector of S $\left(\text{i.e. } \frac{4}{5}\mathbf{u} + \frac{1}{5}\mathbf{t}\right)$, the coefficient of **t** is 1 – (the coefficient of **u**). This would be the case for any point on the line joining T and U. The rock has position vector $\frac{3}{5}\mathbf{u} + \frac{1}{5}\mathbf{t}$. Now, the coefficient of **t** $\left(\text{i.e. } \frac{3}{5}\right)$ should be 1 – (the coefficient of **u**) $\left(\text{i.e. } \frac{1}{5}\right)$, but $1 - \frac{3}{5} = \frac{2}{5}$ so this is not true.
    Hence the rock does not lie on the line joining T and U.

**6** (a)

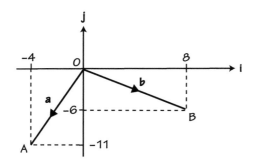

$\mathbf{AB} = \mathbf{AO} + \mathbf{OB}$

$= -\mathbf{a} + \mathbf{b}$

$= \mathbf{b} - \mathbf{a}$

$= (8\mathbf{i} - 6\mathbf{j}) - (-4\mathbf{i} - 11\mathbf{j})$

$= 12\mathbf{i} + 5\mathbf{j}$

(b) Magnitude of $\mathbf{AB} = \sqrt{12^2 + 5^2}$

$= \sqrt{169}$

$= 13$

(c)

> Mark point M on the diagram.

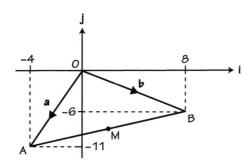

As $\mathbf{m}$ is the position vector of the mid-point M, $\mathbf{OM} = \mathbf{m}$

$\mathbf{BM} = \mathbf{BO} + \mathbf{OM}$

$= -\mathbf{b} + \mathbf{m}$

$= \mathbf{m} - \mathbf{b}$

$\mathbf{MA} = \mathbf{MO} + \mathbf{OA}$

$= -\mathbf{m} + \mathbf{a}$

$= \mathbf{a} - \mathbf{m}$

As M is the mid-point of AB

$\mathbf{BM} = \mathbf{MA}$

$\mathbf{m} - \mathbf{b} = \mathbf{a} - \mathbf{m}$

$2\mathbf{m} = \mathbf{a} + \mathbf{b}$

$\mathbf{m} = \frac{1}{2}(\mathbf{a} + \mathbf{b})$

$= \frac{1}{2}(-4\mathbf{i} - 11\mathbf{j} + 8\mathbf{i} - 6\mathbf{j})$

$= \frac{1}{2}(4\mathbf{i} - 17\mathbf{j})$

**7** (a) $\mathbf{AB} = \mathbf{AO} + \mathbf{OB}$

$\quad\quad = -\mathbf{a} + \mathbf{b} = \mathbf{b} - \mathbf{a}$

$\quad\quad = (14\mathbf{i} - 2\mathbf{j}) - (2\mathbf{i} + 3\mathbf{j})$

$\quad\quad = 12\mathbf{i} - 5\mathbf{j}$

(b) $|\mathbf{AB}| = \sqrt{(12)^2 + (-5)^2} = 13$

(c)

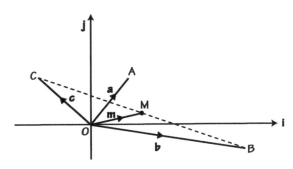

As $\mathbf{m}$ is the position vector of the mid-point M, $\mathbf{OM} = \mathbf{m}$

$\quad \mathbf{OM} = \mathbf{OB} + \mathbf{BM}$

$\quad \mathbf{BM} = \mathbf{OM} - \mathbf{OB}$

$\quad\quad = \mathbf{m} - \mathbf{b}$

Also, $\mathbf{MC} = \mathbf{MO} + \mathbf{OC}$

$\quad\quad\quad = -\mathbf{m} + \mathbf{c}$

$\quad\quad\quad = \mathbf{c} - \mathbf{m}$

Now $\mathbf{BM} = \mathbf{MC}$, so $\mathbf{m} - \mathbf{b} = \mathbf{c} - \mathbf{m}$

$\quad\quad 2\mathbf{m} = \mathbf{b} + \mathbf{c}$

$\quad\quad \mathbf{m} = \frac{1}{2}(\mathbf{b} + \mathbf{c})$

$\quad\quad \mathbf{m} = \frac{1}{2}(-8\mathbf{i} + 3\mathbf{j} + 14\mathbf{i} - 2\mathbf{j}) = 3\mathbf{i} + \frac{1}{2}\mathbf{j}$

> The vectors with $\mathbf{i}$ and $\mathbf{j}$ can now be substituted for $\mathbf{b}$ and $\mathbf{c}$.

(d) As P divides AC in the ratio $3:7$, $\mathbf{AP}:\mathbf{PC} = 3:7$

Hence, $7\mathbf{AP} = 3\mathbf{PC}$

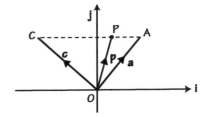

If $\mathbf{p}$ is the position vector $\mathbf{OP}$, then $7(\mathbf{p} - \mathbf{a}) = 3(\mathbf{c} - \mathbf{p})$

$\quad 10\mathbf{p} = 3\mathbf{c} + 7\mathbf{a}$

$\quad\quad\quad = 3(-8\mathbf{i} + 3\mathbf{j}) + 7(2\mathbf{i} + 3\mathbf{j})$

$\quad\quad \mathbf{p} = -\mathbf{i} + 3\mathbf{j}$

# Unit 2  Applied Mathematics A
## Section A: Statistics

## 1  Statistical sampling

**1** (a) The population is all the members of the set being studied/investigated. In this situation, that means the 2000 students who attend the school.

A sample is a smaller subset of the population that is used to make conclusions and inferences about the population. In this situation, that would mean the students consulted by the head teacher.

(b) (i) Attach a different number to the names of each student from 1 to 2000. Generate 200 random numbers from 1 to 2000 using a calculator, computer program or tables and if the number has been selected already, select again until you have 200 numbers.

Match each number to the name of the student and give them the questionnaire to complete.

(ii) An advantage is that, as the sample is truly random, there shouldn't be any sampling bias. A disadvantage is that it is time consuming to generate the sample.

(c) (i) Sampling interval $= \dfrac{\text{population}}{\text{sample size}} = \dfrac{2000}{200} = 10$

(ii) Number each student from 1 to 2000.

Imagine all these numbers are arranged in a circle.

Pick a random number as a point to start.

Calculate the sampling interval (i.e. 10 in this case).

Record the starting number and then every tenth number after that record the number until all 200 numbers have been picked.

Match each number to the name and send them the questionnaire.

(iii) One advantage such as:       The sample is easy to select.

One disadvantage such as:   Less random than simple random sampling.

**2** Give each order a number from 1 to 500.

Set up a random number generator (online or use a calculator) to generate random integers from 1 to 500.

If a random number has already been picked ignore it and pick another.

Continue the process until 50 numbers have been picked.

Using the 50 random numbers, pick out the respective orders for the sample.

**3** (a) Mean $= \dfrac{1+1+3+1+5+4+2+4+3+1}{10} = 2.5$

(b) (i) Sampling interval $= \dfrac{\text{population}}{\text{sample size}} = \dfrac{30}{5} = 6$

(ii) 5, 5, 5, 5, 2

(iii) Mean $= \dfrac{5+5+5+5+2}{5} = 4.4$

> The random number 5 means that you count along to the fifth value in the list. This is then the first data item in the sample. Now count along six numbers and this gives the second data value. This is repeated until the 5 data values are obtained.

(c) The mean obtained using the systematic sampling was higher because the sampling interval just happened to hit on the higher values (i.e. mainly 5s). A different sampling interval would have led to a lower mean.

The opportunity sample used data more reflective of the rest of the data and gave a more accurate mean in this case.

(d) Opportunity sampling:

Advantage – easy to take the sample as it is carried out in the most convenient way.

Disadvantage – sample is not picked at random, so it can be unrepresentative of the population.

Systematic sampling:
Advantage – it is a random sampling method.
Disadvantage – using the sampling interval you can hit on unrepresentative values.

# 2  Data presentation and interpretation

**1**  (a)  Quantitative data is always numerical.                                         True
   (b)  Discrete data can take all values.                                          False
   (c)  Bar charts have quantitative data on the *x*-axis.                          False
   (d)  The height of a bar in a bar chart represents the frequency.               True
   (e)  Histograms have gaps between the bars.                                     False
   (f)  There are numerical values on both axes of a histogram.                    True
   (g)  Histograms always have bars of unequal width.                             False
   (h)  The area of a bar on a histogram represents the frequency density.         False

**2**  (a)  A positively skewed distribution is skewed to the right.                       True
   (b)  A perfectly symmetrical distribution has no skew.                          True
   (c)  A scatter graph with negative correlation has a positive gradient.         False
   (d)  The mode, mean and standard deviation are all measures
      of central tendency.                                                      False
   (e)  The interquartile range is the spread of the middle half of the
      data when the data is arranged in order of size.                          True

**3**  (a)  (i)  Positive correlation
      (ii)  The greater the arm span the greater the height.
   (b)  (i)  The gradient represents the rate of change of arm span with height.
      (ii)  The point at (180, 159) should be circled. Removing this point would lower the
         mean height and raise the mean arm span.
      (iii)  Probably not.
         The arm span of the baby is well outside the collected data, so the line would
         need to be extrapolated. It is not certain that beyond the first and last readings
         the regression equation holds.
      (iv)  Yes, probably causal. Taller people are likely to have other physiological
         measurements in proportion.

**4**  (a)

| Summary statistic | Boys | Girls |
|---|---|---|
| Lower quartile | 30.75 | 36 |
| Median | 46.5 | 57.5 |
| Upper quartile | 61.25 | 65.25 |
| Highest mark | 90 | 85 |
| Lowest mark | 15 | 20 |
| Mean | 47.6 | 52.2 |
| Range | 75 | 65 |
| Inter-quartile range | 30.5 | 29.25 |

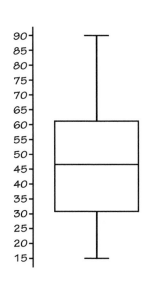

  (b)  See diagram.
  (c)  The range of the marks for the boys is higher but the spread for the middle half of
     the data as shown by the IQR is only slightly higher.
     Both the mean and median are significantly higher for the girls.
     Overall, the girls did better in the exam and their marks were less spread out
     showing they were more consistent.

# 3 Probability

**1** (a) Mutually exclusive events are events that cannot happen together. For example, if event A is 'throw a 6 on the first throw of a die' and event B is 'throw a 3 on the first throw of a die', then event $A$ can happen and event $B$ can happen, but both of them cannot happen.

Independent events are where one of the events happening does not influence the probability of the other happening and vice versa. So, if event A is 'throw a 6 on the first throw of a die' and event B is 'get a head on the first flip of a coin', then if $A$ happens, this will not affect the chance of $B$ happening and vice versa.

(b) $P(A \cup B) = P(A) + P(B) - P(A \cap B)$

$6 \times P(A \cap B) = P(A) + P(B) - P(A \cap B)$

$7 \times P(A \cap B) = P(A) + P(B)$

$7 \times P(A \cap B) = 0.5 + 0.2$

$7 \times P(A \cap B) = 0.7$

$P(A \cap B) = 0.1$

(c) $P(A \cup B) = 6 \times P(A \cap B)$

$= 6 \times 0.1 = 0.6$

**2** (a) $P(A \cup B) = P(A) + P(B)$

$0.4 = 0.25 + P(B)$

$P(B) = 0.15$

> When events $A$ and $B$ are mutually exclusive, there is no overlap between sets $A$ and $B$.

(b) $P(A \cup B) = P(A) + P(B) - P(A \cap B)$

$= P(A) + P(B) - P(A) \times P(B)$

$0.4 = 0.25 + P(B) - 0.25 \times P(B)$

$0.15 = 0.75 \times P(B)$

$P(B) = 0.2$

> Use this equation for independent events.

> Use $P(A \cap B) = P(A) \times P(B)$

> Divide both sides by 0.75

**3** (a) If the events $A$ and $B$ are independent,

$P(A \cup B) = P(A) + P(B) - P(A \cap B)$

$P(A \cap B) = P(A) \times P(B)$

$= 0.2 \times 0.5$

$= 0.1$

$P(A) + P(B) - P(A \cap B) = 0.2 + 0.5 - 0.1 = 0.6$

As $P(A \cup B) \neq 0.6$ the events are not independent.

(b) It means that one event happening influences the probability of the other event happening.

**4** (a) $P(A \cup B) = P(A) + P(B) = 0.35 + 0.45 = 0.8$

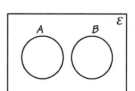

> As events $A$ and $B$ are mutually exclusive $A$ and $B$ cannot both occur so there is no overlap between the sets. Note you are not required to draw any Venn diagrams.

(b) $P(A \cup B) = P(A) + P(B) - P(A \cap B)$

$= 0.35 + 0.45 - (0.35 \times 0.45)$

$= 0.6425$

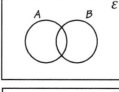

> Notice that the overlap would be added twice if you just consider $P(A) + P(B)$. Hence we need to subtract one of these overlaps, i.e. $P(A \cap B)$.

(c) $P(A \cup B) = P(B) = 0.45$

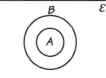

> Set $A$ is included inside set $B$ so $P(A)$ is already included in $P(B)$.

**5** (a) $P(A \cup B) = P(A) + P(B) - P(A \cap B)$

$0.5 = 0.2 + P(B) - P(A \cap B)$

$P(A \cap B) = P(A) \times P(B) = 0.2 \times P(B)$

$0.5 = 0.2 + P(B) - 0.2 \times P(B)$

$0.5 = 0.2 + 0.8\,P(B)$

$0.3 = 0.8\,P(B)$

$P(B) = 0.375$

There are two unknowns in this equation so we need to look for a different equation so that they can be solved simultaneously.

Another equation is created by finding $P(A \cap B)$.

(b) One way to do this is to draw a Venn diagram and mark on it the various probabilities.

$P(A \cap B) = P(A) \times P(B) = 0.2 \times 0.375 = 0.075$

$A$ happens and $B$ doesn't or $B$ happens and $A$ doesn't.

$P(A \text{ only happens}) = 0.125$

$P(B \text{ only happens}) = 0.3$

$P(\text{only } A \text{ or only } B \text{ happens}) = P(A \text{ only happens}) + P(B \text{ only happens})$

$= 0.125 + 0.30$

$= 0.425$

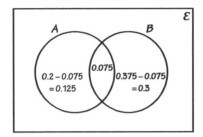

# 4 Statistical distributions

**1** (a) Binomial is not suitable as you do not know the total number of throws, $n$, of the die.

(b) Binomial is not suitable because, when the balls are removed without replacement, it alters the subsequent probability. This means probability $p$ of picking a red ball is not constant which means the events are not independent.

(c) Binomial is suitable as both the probability $p$ and the number of trials $n$ are known. The trials are independent as the probability $p$ stays constant.

(d) Binomial is suitable as both $p$ and $n$ are known and the probability of hitting the bull stays constant which means the throws are independent.

(e) Binomial is not suitable because $p$, the probability of breaking a glass, is unknown.

(f) Binomial is suitable because both $n$ and $p$ are known and the probability of flowering is the same for each bulb.

**2** (a) $n = 10$, $p = 0.09$ and $x = 5$

$P(X = 5) = P(X \leq 5) - P(X \leq 4)$

$= 0.9999 - 0.9990$

$= 0.0009$

The 'Binomial distribution function' tables have been used to answer this question. You could have alternatively used the formula.

(b) $n = 10$, $p = 0.09$ and $x = 0$

$P(X = 0) = 0.3894$

(c) $P(X > 5) = 1 - P(X \leq 5)$

$= 1 - 0.9999$

$= 0.0001$

(d) The trials are independent of each other.

**3** (a) (i) $n = 30$, $p = 0.65$ and $x = 20$

Distribution is $B(20, 0.65)$

$$P(X = x) = \binom{n}{x}p^x(1 - p)^{n-x}$$

$$P(X = 20) = \binom{30}{20}(0.65)^{20}(1 - 0.65)^{30 - 20}$$

$$P(X = 20) = \binom{30}{20}(0.65)^{20}(0.35)^{10}$$

$$= 0.1502$$

(ii) Let the number of bulbs that fail to grow = $Y$

$Y$ is distributed as B(30, 0.35)

$P(X \geq 15) = P(Y \leq 14)$

$\qquad = 0.9348$

(b) Distribution is B($n$, 0.65)

$$P(X = n) = \binom{n}{n}(0.65)^n (1 - 0.65)^{n-n}$$

$$= \binom{n}{n}(0.65)^n (1 - 0.65)^0$$

$$= (0.65)^n$$

$P(X = n) = 0.005688$

So $(0.65)^n = 0.005688$

Taking $\log_e$ of both sides

$\log_e 0.65^n = \log_e 0.005688$

$n\log_e 0.65 = \log_e 0.005688$

$$n = \frac{\log_e 0.005688}{\log_e 0.65} = 12$$

**4** (a) (i) $n = 25$, $p = 0.8$ and $x = 10$

Distribution is B(25, 0.8)

$$P(X = x) = \binom{n}{x}p^x (1 - p)^{n-x}$$

$$P(X = 10) = \binom{25}{10}(0.8)^{10} (1 - 0.8)^{25-10}$$

$$P(X = 10) = \binom{25}{10}(0.8)^{10} (0.2)^{15}$$

$$= 0.0000115$$

(ii) Let $Y = 25 - X$

$Y$ has the distribution B(25, 0.2)

$$P(10 \leq X \leq 15) = P(10 \leq Y \leq 15)$$

$$= P(Y \leq 15) - P(Y \leq 9)$$

$$= 0.9999 - 0.9827$$

$$= 0.0172$$

(b) Mean, $\lambda = 0.04 \times 300 = 12$

$$P(Y = y) = e^{-\lambda}\frac{\lambda^y}{y!}$$

$$P(Y = 5) = e^{-12}\frac{(12)^5}{5!}$$

$$= 0.0127$$

**5** $\lambda = np = 10 \times 0.8 = 8$

As we want to find $P(X < 5)$ the Poisson distribution function tables find the probability of the random variable $X$ with a mean $m$ being less than or equal to $x$.

We therefore find $P(X \leq 4)$ from the table. This gives the cumulative probability that there are fewer than 5 errors (i.e. the total probability of 0, 1, 2, 3 or 4 errors).

$P(X \leq 4) = 0.0996$

> The mean is denoted by the letter m in the tables rather than $\lambda$.

**6** (a) Mean, $\lambda = 0.1 \times 15 = 1.5$

$$P(X = x) = e^{-\lambda}\frac{\lambda^x}{x!}$$

$$P(X = 2) = e^{-1.5}\frac{(1.5)^2}{2!} = 0.2510$$

> This is in the formula booklet so you don't need to remember it.

(b) $P(X > 2) = 1 - [P(X = 2) + P(X = 1) + P(X = 0)]$

$$= 1 - \left[0.2510 + e^{-1.5}\frac{(1.5)^1}{1!} + e^{-1.5}\frac{(1.5)^0}{0!}\right] = 0.1911$$

> 0! and $(1.5)^0$ are both 1.

**7** (a) $P(X = x) = e^{-\lambda}\dfrac{\lambda^x}{x!}$

Mean number of raisins per cookie, $\lambda = \dfrac{400}{100} = 4$

$P(X = 0) = e^{-4}\dfrac{(4)^0}{0!} = 0.0183$

(b) $P(X = 0) = 1\% = 0.01$

We now look at the Poisson distribution function table and look along the rows for $x = 0$ until we find a probability in the body of the table of 0.01 or less. $m = 5$ gives a probability of 0.0067 which is less than 0.01.

Hence the mean number of raisins per cookie needs to be 5.

Now 5 raisins in each of 100 cookies would need $5 \times 100 = 500$ cookies in the batch.

**8** $P(X = x) = e^{-\lambda}\dfrac{\lambda^x}{x!}$

Mean number of phone calls, $\lambda = 2.5$

$P(X = 6) = e^{-2.5}\dfrac{(2.5)^6}{6!} = 0.0278$

> Phone calls arriving, or customers coming into a shop, are classic Poisson 'arrival' problems.

**9** (a) (i) Let $X$ = the number of female dogs.

Distribution is B(20, 0.55)

$$P(X = x) = \binom{n}{x}p^x(1-p)^{n-x}$$

$$P(X = 12) = \binom{20}{12}(0.55)^{12}(1-0.55)^{20-12}$$

$$P(X = 20) = \binom{20}{12}(0.55)^{12}(0.45)^8$$

$$= 0.1623$$

> In part (ii) we are going to use tables so we need to change this to consider the male dogs. This is because if we consider the female dogs we need to use a probability of 0.55 and the tables only go up to 0.50. If you work this out using a calculator, you can still just consider the female dogs.

(ii) Let $Y$ = the number of male dogs

Distribution is B(20, 0.45)

$P(8 \le X \le 16) = P(4 \le Y \le 12)$

$\qquad = P(Y \le 12) - P(Y \le 3)$

$\qquad = 0.9420 - 0.0049 = 0.9371$

(b) Let $R$ = the number of yellow dogs.

$R$ is B(60, 0.05) $\approx$ Po(3)

$\qquad P(R < 5) = P(R \le 4) = 0.8153$

> Poisson distribution tables are used to look this up.

**10** (a) $P(X = x) = e^{-\lambda}\dfrac{\lambda^x}{x!}$

Mean number of emergencies, $\lambda = 8$

$P(X = 7) = e^{-8}\dfrac{(8)^7}{7!}$

$\qquad = 0.1396$

(b) $P(X < 10) = P(X \le 9) = 0.7166$

> As no method is specified, you could also use the tables here.

> The 'Poisson distribution function' table can be used to look up this probability.

# 5 Statistical hypothesis testing

**1** Let the test statistic $X$ = the number of times a goal is scored by the footballer taking a penalty.

The null hypothesis is $\mathbf{H_0}: p = 0.5$

The alternative hypothesis is $\mathbf{H_1}: p > 0.5$

$\mathbf{H_0}$ is B(20, 0.5)

$P(X \ge 15) = 1 - P(X \le 14)$

$\qquad = 1 - 0.9793 = 0.0207$

Now $0.0207 < 0.05$

Hence, we reject the null hypothesis $\mathbf{H_0}$ in favour of the alternative hypothesis $\mathbf{H_1}$.

So, the manager is right in that his scoring penalties probability is greater than 50%.

**2** The null hypothesis $H_0$ is that the coin is fair.

Hence $H_0 : p = 0.5$

$\qquad H_1 : p \neq 0.5$

This is a two-tailed test, so the critical region consists of a region at the end of each tail.

The significance level is divided by two to give the probability at each tail (i.e. 0.05).

$H_0$ is B(10, 0.5)

We consider the tail to the right and find the probability of 7 or more heads occurring.

$$P(X \geq 7) = 1 - P(X < 7)$$
$$= 1 - P(X \leq 6)$$
$$= 1 - 0.8281$$
$$= 0.1719$$

This is not in the critical region as $P(X \geq 7)$ is not less than 0.05.

Hence, the null hypothesis is not rejected at the 10% level of significance.

**3** Let the test statistic $X$ = number of days she sees one or more red squirrels.

The null hypothesis is $H_0 : p = 0.05$

The alternative hypothesis is $H_1 : p \neq 0.05$

$X$ is B(6, 0.05)

As this is a two-tailed test we need to halve the significance level, so this is $\frac{0.01}{2} = 0.005$.

The number of days out of 6 one or more red squirrels would be expected to be seen = $np = 6 \times 0.05 = 0.3$

3 days is greater than this, so the $p$-value will be the probability of $X$ being 3 or more under the null hypothesis.

$$P(X \geq 3) = 1 - P(X \leq 2)$$
$$= 1 - 0.9978$$
$$= 0.0022$$

As $0.0022 \leq 0.005$ the result is significant.

> Make sure that you clearly state the meaning of the result you have obtained.

Hence, there is sufficient evidence at the 1% level of significance to reject the null hypothesis in favour of the alternative hypothesis that the probability of seeing a red squirrel has changed.

**4** Let the test statistic $X$ = the number of exams passed.

The null hypothesis is $H_0 : p = 0.45$

The alternative hypothesis is $H_1 : p < 0.45$

$X$ is B(10, 0.45)

Now as the null hypothesis was rejected, $P(X \leq N) < 0.05$

Tables are used to find the largest value of $N$ so that the probability $< 0.05$

Using tables $N \leq 1$ (i.e. $N$ can take the values of 0 or 1)

**5** (a) (i) The null hypothesis is $H_0 : p = 0.3$

(ii) The alternative hypothesis is $H_1 : p < 0.3$

(iii) The test statistic is $X$ = number of patients waiting more than 30 minutes after their appointment time.

(b) $X$ is B(30, 0.3)

$P(X \leq 4) = 0.0302$ (i.e. 3.02%)

Critical region is $X \leq 4$

> Binomial tables are used to find the largest value of $x$ that gives a probability of less than 0.05 when $n = 30$ and $p = 0.3$.

(c) 6 is not in the critical region so we fail to reject the null hypothesis, so the lead dentist's claim is unjustified according to this evidence.

# Unit 2  Applied Mathematics A
# Section B: Mechanics

## 6  Quantities and units in mechanics

**1**  (a)  $m\,s^{-2}$
   (b)  $m\,s^{-1}$
   (c)  $kg\,m^{-3}$
   (d)  $N$
   (e)  $N$
   (f)  $N\,m$

**2**  $19.32\,g = \dfrac{19.32}{1000}\,kg = 0.01932\,kg$

   $1\,cm^3 = \dfrac{1}{1\,000\,000} = 1 \times 10^{-6}\,m^3$

   Density of gold $= \dfrac{\text{mass in kg}}{\text{volume in m}^3} = \dfrac{0.01932}{1 \times 10^{-6}} = 19\,320\,kg\,m^{-3}$

## 7  Kinematics

**1**  (a)

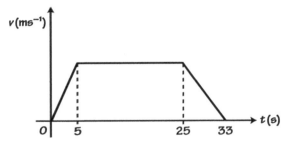

   (b)  $u = 0\,m\,s^{-1}$,      $a = 0.9\,m\,s^{-2}$,      $t = 5\,s$,      $v = ?$
       Using   $v = u + at$
       gives   $v = 0 + 0.9 \times 5 = 4.5\,m\,s^{-1}$
   (c)  Acceleration = gradient of the graph between $t = 25$ and $t = 33\,s$
       $= \dfrac{0 - 4.5}{33 - 25} = -0.56\,m\,s^{-2}$
       Hence deceleration = $0.56\,m\,s^{-2}$
   (d)  Total distance travelled = area under the velocity–time graph
       $= \dfrac{1}{2}(20 + 33) \times 4.5$
       $= 119.25\,m$

> Note that if you say that this is a deceleration, then you must remove the minus sign.

> The area of a trapezium formula,
> $$A = \frac{1}{2}(a + b)h$$
> is used here to work out the distance.

**2**  (a)  Taking the downward direction as positive.
       $u = 0\,m\,s^{-1}$,      $a = g = 9.8\,m\,s^{-2}$,      $s = 160\,m$,      $v = ?$
       Using   $v^2 = u^2 + 2as$ gives
           $v^2 = 0^2 + 2 \times 9.8 \times 160$
           $v = 56\,m\,s^{-1}$

   (b)  Using   $v = u + at$
           $t = \dfrac{v - u}{a} = \dfrac{56 - 0}{9.8} = \dfrac{40}{7}\,s$

(c) The only force acting is gravity; so air resistance is neglected.
(You could also have the answer that the object is modelled as a particle.)

**3** (a) Taking upwards as the positive direction, we have

$u = 15 \text{ m s}^{-1}$,     $a = g = -9.8 \text{ m s}^{-2}$,     $v = 0 \text{ m s}^{-1}$,     $t = ?$

Using     $v = u + at$   gives

$0 = 15 - 9.8t$

Hence,     $t = 1.53 \text{ s}$ (correct to 2 decimal places)

> Always establish a positive direction before you start. Note it does not matter which direction you use. Any quantities that end up negative will be in the opposite direction.

(b) Using     $v^2 = u^2 + 2as$   gives

$0^2 = 15^2 - 2 \times 9.8 \times s$

$s = \dfrac{-225}{-19.6} = 11.48 \text{ m}$ (correct to 2 decimal places)

> At the maximum height, the final velocity, $v$, is zero.

(c) Air resistance has been ignored.

**4** (a) (i)  Constant deceleration
     (ii)  Constant acceleration
     (iii)  Constant deceleration

(b) Displacement in first 20 s = area of triangle OAB

$$= \frac{1}{2} \times 20 \times 15 = 150 \text{ m}$$

Let $t$ = time when the particle returns to the origin again.

Time represented by BD = $(t - 20)$ s

Displacement from $t = 20$ s to $t = t$ s = area of triangle BCD

$$= \frac{1}{2} \times (t - 20) \times 10$$

The two areas are equal, so the displacements are equal in magnitude.

Hence     $150 = \dfrac{1}{2} \times (t - 20)10$

$150 = 5t - 100$

$t = 50 \text{ s}$

> Areas under the time axis represent negative displacements which means the body is moving back towards its starting position.

(c) Average speed $= \dfrac{\text{distance travelled}}{\text{time taken}} = \dfrac{200}{50} = 4 \text{ m s}^{-1}$

**5** Note that this is an unstructured question as no guidance/steps are provided.

Start off by drawing a velocity–time graph. Note there is sometimes more than one method of answering the question. If you look at the velocity–time graph carefully, you may be able to think up an alternative method.

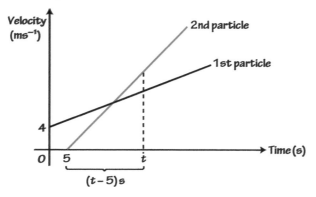

Let $t$ be the time when the 2nd particle overtakes the 1st. At this time, both particles will have travelled the same distance from 0.

Distance travelled by 1st particle $= ut + \frac{1}{2}at^2$

$$= 4t + \frac{1}{2}4t^2$$

$$= 4t + 2t^2 \qquad\qquad (1)$$

Distance travelled by 2nd particle in time $(t - 5)$ $s = ut + \frac{1}{2}at^2$

$$= 0 + \frac{1}{2} \times 10(t - 5)^2$$

$$= 5(t - 5)^2 \qquad (2)$$

Equating the two distances, we obtain:

$$5(t - 5)^2 = 4t + 2t^2$$

$$5(t^2 - 10t + 25) = 4t + 2t^2$$

$$5t^2 - 50t + 125 = 4t + 2t^2$$

$$3t^2 - 54t + 125 = 0$$

Using the formula to solve this quadratic equation we obtain

$$t = \frac{-b \pm \sqrt{b^2 - 4ac}}{2a}$$

$$= \frac{54 \pm \sqrt{(-54)^2 - 4(3)(125)}}{2(3)}$$

$$= 15.27 \text{ s or } 2.73 \text{ s}$$

If you obtain two values always ask yourself if both values are acceptable in the context of the question.

The answer cannot be 2.73 s as this is before the 2nd particle has started.

Time when they have both travelled equal distances = 15.27 s

Substituting this value of $t$ into equation (1), we obtain

$$s = 4t + 2t^2$$

$$= 4(15.27) + 2(15.27)^2$$

$$= 527.43$$

$$= 527 \text{ m (nearest m)}$$

Always check back to the question to see if it specifies a certain accuracy for the answer.

**6** (a)

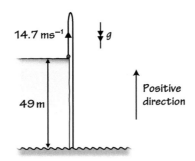

Taking the upward direction as positive.

$u = 14.7 \text{ m s}^{-1}$, $\quad v = ?$, $\quad s = -49 \text{ m}$ $\quad a = -9.8 \text{ m s}^{-2}$,

$$s = ut + \frac{1}{2}at^2$$

$$-49 = 14.7t + \frac{1}{2} \times (-9.8)t^2$$

$$4.9t^2 - 14.7t - 49 = 0$$

Dividing this equation by 4.9 we obtain

$$t^2 - 3t - 10 = 0$$

$$(t - 5)(t + 2) = 0$$

Hence, $t = 5$ s ($t = -2$ s is an impossible time so this is ignored).

Always check to see if the equation can be simplified by dividing through by a number. This will make factorising easier. If you are unable to factorise it, then use the formula instead.

(b) Taking the upward direction as positive.

$$v = u + at$$

$$= 14.7 - 9.8 \times 5$$

$$= -34.3 \text{ m s}^{-1}$$

(Note that the negative sign shows this velocity is in the opposite to the direction taken as positive so this velocity is downwards)

Hence speed of stone before it hits the sea = 34.3 m s$^{-1}$

As speed is a scalar it has size only so don't insert the minus sign.

**7** (a) $v = 4 + 3t - t^2$

When $v = 0$, $\qquad 4 + 3t - t^2 = 0$

Factorising gives $(4 - t)(1 + t) = 0$

The positive solution is $t = 4\,\text{s}$

$$a = \frac{dv}{dt} = 3 - 2t$$

When $t = 4\,\text{s}$, $\qquad a = \frac{dv}{dt} = 3 - 2(4) = -5\,\text{m s}^{-2}$

> To differentiate, you multiply by the index and then reduce the index by 1.

(b) $r = \int v\,dt = \int(4 + 3t - t^2)dt = 4t + \dfrac{3t^2}{2} - \dfrac{t^3}{3} + c$

When $t = 0\,\text{s}$, $r = 0$ putting these into the above equation gives $c = 0$.

When $t = 4\,\text{s}$, $r = 4(4) + \dfrac{3(4)^2}{2} - \dfrac{(4)^3}{3} = 16 + 24 - \dfrac{64}{3} = 18\dfrac{2}{3}$

Average velocity $= \dfrac{\text{total distance travelled}}{\text{time taken}} = \dfrac{18\frac{2}{3}}{4} = 4.7\,\text{m s}^{-1}$

> To integrate you increase the index by 1 and then divide by the new index. Remember to include a constant.

**8** (a) $v = 8 + 7t - t^2$

$\qquad\qquad t^2 - 7t - 8 = 0$

$\qquad\qquad (t + 1)(t - 8) = 0$

$t = -1\,\text{s}$ (which is impossible) or $t = 8\,\text{s}$

Hence time $= 8\,\text{s}$

(b) $v = 8 + 7(0) - (0) = 8\,\text{m s}^{-1}$

(c) $t = 0$ to $t = 1$ is the 1st second

$t = 1$ to $t = 2$ is the 2nd second

$r = \int_1^2 v\,dt = \int_1^2 (8 + 7t - t^2)dt$

$\quad = \left[8t + \dfrac{7t^2}{2} - \dfrac{t^3}{3}\right]_1^2$

$\quad = \left[8(2) + \dfrac{7(2)^2}{2} - \dfrac{(2)^3}{3} - \left(8 + \dfrac{7}{2} - \dfrac{1}{3}\right)\right]$

$\quad = 16\dfrac{1}{6}\,\text{m}$

> You could use the alternative method involving finding the indefinite integral and the constant of integration. You could then use this to find the distances for $t = 1\,\text{s}$ and $t = 2\,\text{s}$. You can then subtract the distances to find the distance travelled in the 2nd second.

> A mixed number means a whole number and a fraction.

**9** (a) $s = \int v\,dt$

$\quad = \int(12t - 3t^2)\,dt$

$\quad = \dfrac{12t^2}{2} - \dfrac{3t^3}{3} + c$

$\quad = 6t^2 - t^3 + c$

$s = 0$ when $t = 1$, hence $0 = 6(1)^2 - (1)^3 + c$, giving $c = -5$

Hence $s = 6t^2 - t^3 - 5$

(b) $a = \dfrac{dv}{dt}$

$\quad = 12 - 6t$

**10** Taking the upward direction as positive

$u = 7\,\text{m s}^{-1}$, $\qquad g = -9.8\,\text{m s}^{-1}$, $\qquad t = 4\,\text{s}$, $\qquad s = ?$

$s = ut + \dfrac{1}{2}at^2$

$\quad = 7 \times 4 + \dfrac{1}{2}(-9.8)4^2$

$\quad = -50.4\,\text{m}$ (as this is a negative displacement it is below the point of projection)

Height of cliff $= 50.4\,\text{m}$

# 8 Dynamics of a particle

**1**

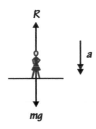

Applying Newton's 2nd law, we have

$$ma = mg - R$$
$$58 \times 2.5 = 58 \times 9.8 - R$$
$$145 = 568.4 - R$$
$$R = 423.4 \, \text{N}$$

> There is a resultant force of *ma* acting in the direction of the acceleration. This resultant force means the weight acting down is larger than the tension acting up.

> The resultant force (i.e. the accelerating force) is provided by the weight minus the reaction.

**2** (a) For the 1st part of the journey if the acceleration is $4 \, \text{m s}^{-2}$ then the velocity increases by $4 \, \text{m s}^{-1}$ each second so if a final speed of $12 \, \text{m s}^{-1}$ is reached then this will have taken 3 s. We need this so we can add the values to the time axis.

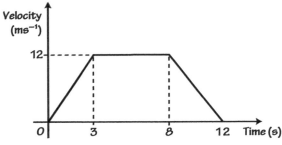

$$\text{Deceleration} = \frac{\text{change in speed}}{\text{time}} = \frac{12}{4} = 3 \, \text{m s}^{-2}$$

> The gradient of the graph between $t = 8 \, \text{s}$ and $t = 12 \, \text{s}$ is the deceleration.

(b) Lift accelerating

Applying Newton's 2nd law of motion, we have

$$ma = R - mg$$
$$50 \times 4 = R - 50 \times 9.8$$
$$R = 690 \, \text{N}$$

Lift travelling with constant speed.

At constant speed there is no resultant acceleration, so the normal reaction is equal to the weight of the man.

$$R = mg = 50 \times 9.8 = 490 \, \text{N}$$

Lift decelerating

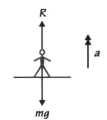

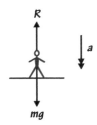

Applying Newton's 2nd law of motion, we have

$$ma = mg - R$$
$$50 \times 3 = 50 \times 9.8 - R$$
$$R = 340 \, \text{N}$$

**3**

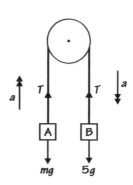

Taking the direction to the left as positive.
Applying Newton's 2nd law of motion to A

$$5a = 126 - T \qquad (1)$$

Applying Newton's 2nd law of motion to B

$$9a = T - 9g \qquad (2)$$

Adding (1) and (2) we obtain

$$14a = 126 - 9g$$
$$a = 2.7 \text{ m s}^{-2}$$

Substituting $a = 2.7$ into equation (1)

$$5 \times 2.7 = 126 - T$$
$$T = 112.5 \text{ N}$$

**4**  (a)

Applying Newton's 2nd law to the 5 kg mass, we have

$$5a = 5g - T$$

As $a = 0.2g$, $\qquad g = 5g - T$

$$T = 4g$$

(b)  Applying Newton's 2nd law to the $m$ kg mass, we have

$$ma = T - mg$$

As $a = 0.2g$ and $T = 4g$, $\qquad 0.2mg = 4g - mg$

$$1.2mg = 4g$$
$$m = \frac{4}{1.2} = 3.3 \text{ kg}$$

(c)  The smooth pulley means there are no frictional forces to take into account.
The light inextensible string means the tension and acceleration stay constant throughout the motion of the masses.

# 9 Vectors

**1**

| Quantity | Scalar | Vector |
|---|---|---|
| Velocity | | ✓ |
| Speed | ✓ | |
| Force | | ✓ |
| Distance | ✓ | |
| Acceleration | | ✓ |
| Displacement | | ✓ |

**2** (a) $\mathbf{R} = \mathbf{P} + \mathbf{S}$
$= (3\mathbf{i} - 20\mathbf{j}) + (-8\mathbf{i} + 8\mathbf{j})$
$= -5\mathbf{i} - 12\mathbf{j}$

(b) $|\mathbf{r}| = \sqrt{(-5)^2 + (-12)^2}$
$= 13\,\text{N}$

(c)
$$\tan \theta = \frac{12}{5}$$
$$\theta = \tan^{-1}\left(\frac{12}{5}\right)$$
$$= 67.4° \ (1\,\text{d.p.})$$

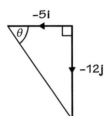

(d) $\mathbf{a} = \dfrac{\mathbf{R}}{m}$
$$= \frac{-5\mathbf{i} - 12\mathbf{j}}{5}$$
$$\mathbf{a} = -\mathbf{i} - 2.4\mathbf{j}$$

> Draw a diagram showing the components of the vector in the horizontal and vertical directions.

> Notice the question is asking for the vector for the acceleration and not the magnitude. Here we use the equation except here the force is the vector **R** and the acceleration is the vector **a**.

**3** (a) Distance travelled from P to Q $= \sqrt{5^2 + 2^2} = \sqrt{29}$
Distance travelled from Q to R $= \sqrt{(-3)^2 + 3^2} = \sqrt{18}$
Total distance travelled from P to R $= \sqrt{29} + \sqrt{18} = 9.63\,\text{km} \ (2\,\text{d.p.})$

(b) $\overrightarrow{PR} = \overrightarrow{PQ} + \overrightarrow{QR}$
$= 5\mathbf{i} + 2\mathbf{j} + -3\mathbf{i} + 3\mathbf{j}$
$= 2\mathbf{i} + 5\mathbf{j}$
$\theta = \tan^{-1}\left(\dfrac{5}{2}\right)$
$\theta = 68.2° \ (1\,\text{d.p.})$

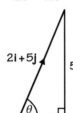

# Sample Test Paper Unit 1
# Pure Mathematics A

**1** (a) $(a + b)^n = a^n + \binom{n}{1}a^{n-1}b + \binom{n}{2}a^{n-2}b^2 + \binom{n}{3}a^{n-3}b^3 + \ldots$

$(a + b)^4 = a^4 + \binom{4}{1}a^3b + \binom{4}{2}a^2b^2 + \binom{4}{3}ab^3 + \binom{4}{4}b^4$

Finding $\binom{4}{1}, \binom{4}{2}, \binom{4}{3}, \binom{4}{4}$ by using the formula or by using Pascal's triangle and substituting them in to the above formula gives:

$(a + b)^4 = a^4 + 4a^3b + 6a^2b^2 + 4ab^3 + b^4$

$\left(x - \dfrac{1}{x}\right)^4 = (x)^4 + 4(x)^3\left(-\dfrac{1}{x}\right) + 6(x)^2\left(-\dfrac{1}{x}\right)^2 + 4(x)\left(-\dfrac{1}{x}\right)^3 + \left(-\dfrac{1}{x}\right)^4$

$= x^4 - 4x^2 + 6 - \dfrac{4}{x^2} + \dfrac{1}{x^4}$

(b) If $x = 1$ is substituted into $\left(x - \dfrac{1}{x}\right)^4$ we obtain $\left(1 - \dfrac{1}{1}\right)^4 = 0$.

If $x = 1$ is substituted into the expansion $x^4 - 4x^2 + 6 - \dfrac{4}{x^2} + \dfrac{1}{x^4}$ we should obtain 0.

Hence $x^4 - 4x^2 + 6 - \dfrac{4}{x^2} + \dfrac{1}{x^4} = 1 - 4 + 6 - 4 + 1 = 0$

The two answers are the same, so the expansion is likely to be correct.

**2** $\dfrac{7}{2\sqrt{14}} + \left(\dfrac{\sqrt{14}}{2}\right)^3 = \dfrac{7\sqrt{14}}{2\sqrt{14}\,\sqrt{14}} + \dfrac{14\sqrt{14}}{8}$

> Any surds in the denominator of terms in the expression need to be removed.

$= \dfrac{7\sqrt{14}}{28} + \dfrac{7\sqrt{14}}{4}$

$= \dfrac{\sqrt{14}}{4} + \dfrac{7\sqrt{14}}{4}$

$= 2\sqrt{14}$

**3** Suppose $f(x) = x^2 + 1$ and $g(x) = x^2 - 2$ then their derivatives are

$f'(x) = 2x \quad \text{and} \quad g'(x) = 2x$

These two derivatives are equal but the original functions are not equal.
Hence the statement given is false.

> Remember that to differentiate a function you multiply by the index and then reduce the index by one.

**4** (a) (i) $y = a^x$

(ii) $a = y^{\frac{1}{x}}$

(iii) $\log_a y^3 = 3\log_a y$

Now $x = \log_a y = 2$

Hence $\log_a y^3 = 3 \times 2 = 6$

(iv) $\log_a (ay)^3 = 3\log_a ay = 3\log_a a + 3\log_a y$

Now $\log_a a = 1 \quad$ and $\quad \log_a y = 2$

Hence, $\log_a (ay)^3 = 3 + 3 \times 2 = 9$

(v) $\log_a\left(\dfrac{y^5}{a^4}\right) = \log_a y^5 - \log_a a^4 = 5\log_a y - 4\log_a a = 10 - 4 = 6$

(b) $2^{x-1} = 3^{(x+3)}$

Taking logs (base 10 here but you could use any other)

$(x - 1)\log_{10} 2 = (x + 3)\log_{10} 3$

$x\log_{10} 2 - \log_{10} 2 = x\log_{10} 3 + 3\log_{10} 3$

$x\log_{10} 2 - x\log_{10} 3 = 3\log_{10} 3 + \log_{10} 2$

$x(\log_{10} 2 - \log_{10} 3) = 3\log_{10} 3 + \log_{10} 2$

$x = \dfrac{3\log_{10} 3 + \log_{10} 2}{\log_{10} 2 - \log_{10} 3}$

**5** (a)
$$x^2 + y^2 - 18x - 22y + 177 = 0$$
$$(x - 9)^2 + (y - 11)^2 - 81 - 121 + 177 = 0$$
$$(x - 9)^2 + (y - 11)^2 = 25$$

Coordinates of P are (9, 11)

Radius $= \sqrt{25} = 5$

> Start off by completing the square for $x$ and $y$.

(b) (i) If point T lies on the circle, its coordinates will satisfy the equation of the circle.

Substituting $x = 5$, $y = 8$ into the equation of C gives
$$(x - 9)^2 + (y - 11)^2 = (5 - 9)^2 + (8 - 11)^2$$
$$= (-4)^2 + (-3)^2$$
$$= 25$$

This is the same as the right-hand side of the equation for C, proving that T lies on the circle.

(ii) Finding the gradient of the radius passing through T (i.e. PT).

Gradient of line joining P(9, 11) and T(5, 8) $= \dfrac{8 - 11}{5 - 9} = \dfrac{-3}{-4} = \dfrac{3}{4}$

Line PT will be a normal to the tangent to the circle at T. Hence the product of the gradients will be −1.

Gradient of tangent at T $= -\dfrac{4}{3}$

Equation of the tangent at T having gradient $-\dfrac{4}{3}$ and passing through the point T(5, 8) is
$$y - 8 = -\frac{4}{3}(x - 5)$$
$$3y - 24 = -4x + 20$$
$$3y + 4x - 44 = 0$$

**6** First find the gradient
$$f'(x) = \frac{3x^2}{3} + \frac{2x}{2} - 12$$
$$= x^2 + x - 12$$

Now we need to find the values of $x$ which will make this expression greater than zero. First find the roots of the equation.
$$x^2 + x - 12 = 0$$
$$(x + 4)(x - 3) = 0$$

From the graph we want those sections of the curve that are above (but not on) the $x$-axis. Hence, range of values for which $x$ is an increasing function is
$$x < -4 \quad \text{or} \quad x > 3$$

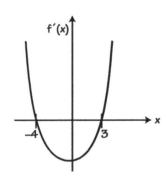

**7** (a) $\overrightarrow{AC} = \overrightarrow{AB} + \overrightarrow{BC}$
$$= (240\mathbf{i} - 60\mathbf{j}) + (-180\mathbf{i} + 200\mathbf{j})$$
$$= 60\mathbf{i} + 140\mathbf{j}$$

> Don't bother working out the square roots of these as you will need to square them later when the cosine rule is used.

(b) $|\overrightarrow{AB}| = \sqrt{240^2 + 60^2} = \sqrt{61\,200}$

$|\overrightarrow{BC}| = \sqrt{180^2 + 200^2} = \sqrt{72\,400}$

$|\overrightarrow{AC}| = \sqrt{60^2 + 140^2} = \sqrt{23\,200}$

Using the cosine rule: $a^2 = b^2 + c^2 - 2bc \cos BAC$
$$72\,400 = 23\,200 + 61\,200 - 2 \times \sqrt{23\,200} \times \sqrt{61\,200} \times \cos BAC$$
$$\cos BAC = 0.1592$$

Angle BAC = 80.8° (nearest 0.1°)

(c) Area of triangle $= \dfrac{1}{2} bc \sin A$
$$= \frac{1}{2} \times \sqrt{23\,200} \times \sqrt{61\,200} \times \sin 80.8$$
$$= 18\,598 \text{ m}^2$$

**8**    For no real roots $b^2 - 4ac < 0$

List the values of $a$, $b$ and $c$ before you start.

$a = m - 1$

$b = 2m$

$c = 7m - 4$

$b^2 - 4ac < 0$

$$(2m)^2 - 4(m - 1)(7m - 4) < 0$$

$$4m^2 - 4(7m^2 - 11m + 4) < 0$$

$$4m^2 - 28m^2 + 44m - 16 < 0$$

$$-24m^2 + 44m - 16 < 0$$

Multiplying through by $-1$, remembering to reverse the inequality sign

$$24m^2 - 44m + 16 > 0$$

Dividing both sides by 4 gives

$$6m^2 - 11m + 4 > 0$$

Put $6m^2 - 11m + 4 = 0$

$$(3m - 4)(2m - 1) = 0$$

$$x = \frac{4}{3} \quad \text{or} \quad \frac{1}{2}$$

If a graph of $(3m - 4)(2m - 1)$ is plotted against $m$ on the $x$-axis the curve is U-shaped, cutting the $x$-axis at $m = \frac{4}{3}$ and $m = \frac{1}{2}$.

The region needed is above the $x$-axis.

Hence the required range of $m$ is $m < \frac{1}{2}$ or $m > \frac{4}{3}$.

**9**    (a)   (i)   Note that the original curve has been translated by $\begin{pmatrix} -2 \\ 0 \end{pmatrix}$

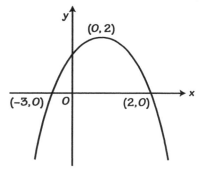

     (ii)   The curve has been reflected in the $x$-axis and stretched by a scale factor 2 parallel to the $y$-axis.

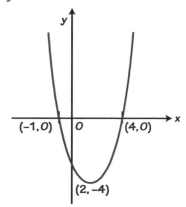

   (b)   The graph of $f(x) = -2f(x) + 4$ is a translation of the graph $f(x) = -2f(x)$ by $\begin{pmatrix} 0 \\ 4 \end{pmatrix}$ and this will translate the point $(2, -4)$ to $(2, 0)$.

      Hence both functions have $x = 2$ as a root.

**10** (a) Let angle ACB = $\theta$

Using the sine rule

$$\frac{16}{\sin\theta} = \frac{8}{\sin 20°}$$

$$\sin\theta = \frac{16\sin 20°}{8}$$

$$\sin\theta = 0.6840$$

$\theta = 43.16°$ or $(180 - 43.16)° = 136.84°$

$\theta = 43.2°$ or $136.8°$

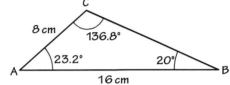
Remember to give your answers to the nearest 0.1°.

(b) Angle BAC = $\big(180 - (43.2 + 20)\big) = 116.8°$ or $\big(180 - (136.8 + 20)\big) = 23.2°$.

We now need to find the side opposite the angle in each of these two situations. The cosine rule could be used here but it is easier to use the sine rule.

For angle BAC = 23.2° and using the sine rule, we have

$$\frac{BC}{\sin 23.2°} = \frac{8}{\sin 20°}$$

$$BC = \frac{8\sin 23.2°}{\sin 20°}$$

$$BC = 9.2145$$

$$= 9.21 \text{ cm (3 significant figures)}$$

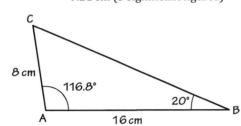

For angle BAC = 116.8° and using the sine rule, we have $\dfrac{BC}{\sin 116.8°} = \dfrac{8}{\sin 20°}$

$$BC = \frac{8\sin 116.8°}{\sin 20°}$$

$$BC = 20.8780$$

$$= 20.9 \text{ cm (3 significant figures)}$$

**11** (a) When $t = 0$, $e^{-kt} = e^0 = 1$, so $N = A$ which means that $A$ is number of radioactive nuclei initially present.

(b) Substituting the pairs of numbers into the equation, we obtain

$$1000 = Ae^{-4k}$$

$$300 = Ae^{-8k}$$

Dividing these two equations to eliminate $A$, we obtain

$$\frac{300}{1000} = \frac{e^{-8k}}{e^{-4k}}$$

$$0.3 = e^{-8k + 4k}$$

$$0.3 = e^{-4k}$$

Taking ln of both sides, we obtain

$$\ln 0.3 = -4k$$

$$k = 0.301 \text{ (3 d.p.)}$$

(c) Substituting $k$ into $1000 = Ae^{-4k}$ to find the value of $A$, we obtain:

$$1000 = Ae^{-4 \times 0.301}$$

Giving $\quad A = 3333$

Now $\quad N = 3333e^{-0.301t}$

When $t = 10$, $\quad N = 3333e^{-0.301 \times 10}$

Giving $\quad N = 164$

(d) $200 = 3333e^{-0.301t}$

$$\frac{200}{3333} = e^{-0.301t}$$
$$0.060006 = e^{-0.301t}$$

Taking ln of both sides:
$$\ln 0.060006 = -0.301t$$
$$t = 9\,\text{s (nearest second)}$$

⑫ Putting $x + \delta x$ and $y + \delta y$ into the equation we have:
$$y + \delta y = 20(x + \delta x)^2 + 9(x + \delta x) - 20$$
$$y + \delta y = 20(x^2 + 2x\delta x + (\delta x)^2) + 9x + 9\delta x - 20$$
$$y + \delta y = 20x^2 + 40x\delta x + 20(\delta x)^2 + 9x + 9\delta x - 20$$
But $\qquad y = 20x^2 + 9x - 20$

Subtracting these equations gives
$$\delta y = 40x\delta x + 20(\delta x)^2 + 9\delta x$$
$$\frac{\delta y}{\delta x} = 40x + 20\delta x + 9$$

Dividing both sides by $\delta x$ and letting $\delta x \rightarrow 0$
$$\frac{dy}{dx} = \lim_{\delta x \to 0}\frac{\delta y}{\delta x} = 40x + 9$$
$$\frac{dy}{dx} = 40x + 9$$

⑬ (a) $y = x^2 - 4x - 5$
$$0 = (x - 5)(x + 1)$$
$$x = 5 \text{ or } -1$$
Hence, A is $(-1, 0)$ and B is $(5, 0)$

(b) $y = x^2 - 4x - 5$
$$\frac{dy}{dx} = 2x - 4$$

> Need to find the gradient at the point D ( 4, –5).

When $x = 4$, $\frac{dy}{dx} = 2(4) - 4 = 4$

Equation of the tangent at D having gradient 4 and passing through the point D(4, –5) is
$$y + 5 = 4(x - 4)$$
$$y = 4x - 21$$

(c) Finding the coordinates of C where the tangent cuts the $x$-axis.
$$0 = 4x - 21$$
$$x = \frac{21}{4} = 5\tfrac{1}{4}$$

Hence, C is $\left(5\tfrac{1}{4}, 0\right)$

We now find the area of the region bounded by the curve and the line $x = 4$ and point B(5, 0).

Area between curve and $x$-axis $= \int_4^5 y\,dx = \int_4^5 (x^2 - 4x - 5)dx = \left[\frac{x^3}{3} - 2x^2 - 5x\right]_4^5$
$$= \left[\left(\frac{125}{3} - 50 - 25\right) - \left(\frac{64}{3} - 32 - 20\right)\right]$$
$$= -\frac{8}{3}$$

The result is negative because the area is below the $x$-axis, so area $= \frac{8}{3}$.

Let E be the point (4, 0), directly above D on the $x$-axis.

Length CE $= 5\tfrac{1}{4} - 4 = 1\tfrac{1}{4}$

Area of triangle CDE $= \frac{1}{2} \times 1\tfrac{1}{4} \times 5 = 3\tfrac{1}{8}$

Shaded area $= 3\tfrac{1}{8} - \frac{8}{3} = \frac{11}{24}$

⑭ Firstly we need to find **the centre of the circle and its radius.**

$$x^2 + y^2 - 8x + 10y + 28 = 0$$
$$(x - 4)^2 + (y + 5)^2 - 16 - 25 + 28 = 0$$
$$(x - 4)^2 + (y + 5)^2 = 13$$

Hence, centre is at $(4, -5)$ and radius is $\sqrt{13}$

Rearranging the equation to find the gradient of the line we obtain

$y = -\frac{2}{3}x + 2$ so the gradient of this line is $-\frac{2}{3}$.

Gradient of the radius which would be at right-angles to the line $= \frac{3}{2}$

Equation of the radius with gradient $\frac{3}{2}$ and passing through the centre $(4, -5)$ is

$$y - y_1 = m(x - x_1)$$
$$y + 5 = \frac{3}{2}(x - 4)$$
$$2y + 10 = 3x - 12$$
$$2y = 3x - 22$$

Solving this equation simultaneously with the equation of the line gives the coordinates of the point of intersection as $(6, -2)$.

If this point of intersection lies on the circumference of the circle, then the line must be a tangent. We can find the distance from the centre $(4, -5)$ to the point of intersection $(6, -2)$ and prove that it is equal to the radius.

Using the formula for the distance between two points we obtain:

$$d = \sqrt{(x_2 - x_1)^2 + (y_2 - y_1)^2}$$
$$= \sqrt{(6 - 4)^2 + (-2 + 5)^2}$$
$$= \sqrt{13}, \text{ which is the same as the radius of the circle.}$$

Hence the line is a tangent to the circle.

⑮ (a) Area of glass $= 2x^2 + 6xh$
$$2x^2 + 6xh = 60\,000$$
$$x^2 + 3xh = 30\,000$$
$$3xh = 30\,000 - x^2$$
$$h = \frac{30\,000 - x^2}{3x}$$

(b) Volume, $V = 2x^2h$
$$V = 2x^2 \frac{30\,000 - x^2}{3x}$$
$$= \frac{60\,000x^2 - 2x^4}{3x}$$
$$= 20\,000x - \frac{2}{3}x^3$$

(c) $\frac{dV}{dx} = 20000 - 2x^2$

$$\frac{dV}{dx} = 0$$
$$20000 - 2x^2 = 0$$
$$x^2 = 10000$$
$$x = \sqrt{10000}$$
$$= 100 \text{ cm}$$
$$\frac{d^2V}{dx^2} = -4x$$

As $x$ cannot be negative or zero, $-4x$ is always negative.

Hence this value of $x$ gives a maximum value of $V$.

Another way to complete this question is to solve the equation of the line and the equation of the circle simultaneously. If the resulting equation has only one real root, then the line must be a tangent to the circle.

# Sample Test Paper Unit 2
## Applied Mathematics A

### Section A – Statistics

**1** (a) (i) A method of sampling that uses the most convenient way of collecting the sample.

(ii) Mean $= \dfrac{0 + 2 + 1 + 4 + 2 + 1 + 5 + 6 + 3 + 5}{10} = 2.9$

(b) (i) Sampling interval $= \dfrac{\text{population}}{\text{sample size}} = \dfrac{25}{5} = 5$

(ii) 4, 3, 5, 6, 5

(iii) Mean $= \dfrac{4 + 3 + 5 + 6 + 5}{5} = 4.6$

(c) The systematic sample because it is a random sample and uses numbers throughout the distribution rather than the first 10 values which may not be typical of the rest of the distribution.

**2** (a) $P(A \cup B) = P(A) + P(B) - P(A \cap B)$ and $P(A \cap B) = P(A) \times P(B)$

So
$$0.4 = 0.2 + P(B) - 0.2 \times P(B)$$
$$0.2 = 0.8 \times P(B)$$
$$P(B) = 0.25$$

(b) $P(A \cap B) = P(A) \times P(B)$
$$= 0.2 \times 0.25 = 0.05$$

> Once P(A ∩ B) has been found, the Venn diagram can be drawn.

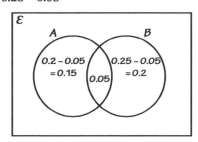

The probability that exactly one event occurs

= probability of only $A$ occurring + probability of only $B$ occurring
$$= 0.15 + 0.2$$
$$= 0.35$$

**3** (a) (i) An outlier is an unrepresentative value. It is not used to produce the box and whisker diagram but is shown as a separate point.

(ii) As the outlier was much higher than all other values, the mean will go down after the outlier is removed.

(iii) The standard deviation is a measure of central variation, removal of the outlier will reduce the variation which will decrease the standard deviation.

(b) (i) College A: Range = 38      IQR = 10.5

College B: Range = 35      IQR = 10

(ii) On average the college A students were taller as shown by a higher mean and median. College B students showed less variability in their height as shown by lower range, IQR and standard deviation.

Both college A and college B students had a roughly symmetrical distribution to their heights so there is almost no skew.

> Note if the mean and median are approximately the same, there is no skew and the distribution is symmetrical. If the mean is greater than the median it is positively skewed and if it less it is negatively skewed.

**4** (a) The number of trials is not known so binomial cannot be used.

The mean can be found, so Poisson is appropriate.

(b)  (i)   Mean = 0.5 × 30 = 15

$$P(X = x) = e^{-\lambda}\frac{\lambda^x}{x!}$$

$$P(X = 18) = e^{-15}\frac{(15)^{18}}{18!}$$

$$= 0.0706$$

(ii)  $P(X > 20) = 1 - P(X \leq 20)$

$$= 1 - 0.9170$$

$$= 0.083$$

> Be careful here. The probability of greater than 20 does not include 20 itself.

**5**  (a)  $\mathbf{H}_0 : p = 0.3$
$\mathbf{H}_1 : p > 0.3$

(b)  As $\mathbf{H}_1 : p > 0.3$ we use the upper tail of the probability distribution.

Assuming the null hypothesis is true, we use B(40, 0.3) to find values of $X$ with a probability that just exceeds 0.95.

From the table $P(X \leq 17) = 0.9680$. The critical value is one more than this value, so the critical value is 18.

Hence the critical region is $X \geq 18$

(c)  $P(X \leq 17) = 0.9680$
$P(X \geq 18) = 1 - 0.9680$

$$= 0.032$$

Hence, actual significance level of the test = 3.2%

**6**  (a)  (i)   Positive correlation.

(ii)  A higher English mark suggests a higher Mathematics mark.

(b)  (i)   Each additional English mark corresponds to an increase in the Mathematics mark of 0.9 marks on average.

(ii)  Yes it would. The positive correlation looks fairly close but it is not as close for lower marks. It would be less accurate for English marks less than 20% as there are no values in that area of the graph, so extrapolation would need to be used.

(iii)  Non-causal. You cannot say that being good at English causes being good at Maths, or vice versa. Both are probably caused by intelligence, which is a separate thing.

## Section B – Mechanics

**1**  (a)  Taking the downward direction as positive, we have the following:

$u = -2\,\text{m s}^{-1}, \quad g = 9.8\,\text{m s}^{-2}, \quad s = 50\,\text{m}, \quad t = ?$

$$v^2 = u^2 + 2as$$
$$v^2 = (-2)^2 + 2 \times 9.8 \times 50$$
$$v^2 = 984$$
$$v = 31.4\,\text{m s}^{-1}$$

Now using $s = \frac{1}{2}(u + v)t$

$$50 = \frac{1}{2}(-2 + 31.4)\,t$$

$$t = 3.4\,\text{s}$$

> You could alternatively use the equation
> $$s = ut + \frac{1}{2}at^2$$
> and solve for $t$ but this will produce a quadratic equation in $t$ which you will have to solve using the formula. This method involves finding $v$ first before finding $t$ is a bit simpler.

(b)  Taking the upward direction as positive

$u = 2\,\text{m s}^{-1}, \quad a = 2\,\text{m s}^{-2}, \quad t = 4\,\text{s}, \quad s = ?$

$$s = ut + \frac{1}{2}at^2$$

$$= 2 \times 4 + \frac{1}{2} \times 2 \times 4^2$$

$$= 24\,\text{m}$$

Total distance above the ground = 50 + 24 = 74 m

**2**    (a)   (i)   $v = 3t^2 + 12$

When $t = 0$ s, $v = 0 + 12 = 12$ m s$^{-1}$

(ii)   The shape of the graph is a parabola with the minimum point at $(0, 12)$.

The graph never goes below the $x$-axis, which means that the velocity is never negative.

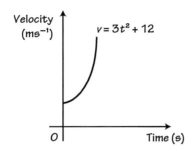

(b)   (i)   The equation $v = 13t + 8$ is of the form $y = mx + c$ and is therefore a straight line with a constant gradient. The gradient of the velocity–time graph represents the acceleration, so the acceleration is constant.

(ii)   Equating the velocities of the two particles we obtain

$$3t^2 + 12 = 13t + 8$$
$$3t^2 - 13t + 4 = 0$$

Factorising this quadratic, we obtain

$$(3t - 1)(t - 4) = 0$$

Solving gives $t = \frac{1}{3}$ or 4 seconds

Hence, they have the same velocity at $t = \frac{1}{3}$ or 4 seconds

> An alternative method would be to differentiate the velocity and show that the result did not contain any terms in $t$ thus showing the acceleration is independent of the time.

**3**    (a)

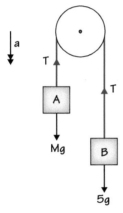

Applying Newton's 2nd law to the heavier particle (i.e. particle A)

$$Ma = Mg - T \qquad\qquad (1)$$

Applying Newton's 2nd law to the lighter particle

$$5a = T - 5g \qquad\qquad (2)$$

Adding (1) and (2)

$$Ma + 5a = Mg - 5g$$
$$a(M + 5) = g(M - 5)$$
$$a = \frac{g(M - 5)}{M + 5}$$

(b)   Because the string is light and inextensible, it is assumed that the tension is constant throughout the string. Because the pulley is smooth, it is assumed that there are no frictional forces acting on the string.

**4** (a)

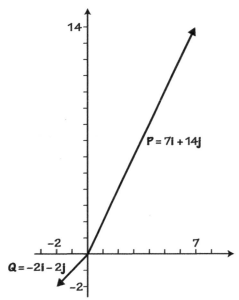

(b)
$$\mathbf{R} = (7\mathbf{i} + 14\mathbf{j}) + (-2\mathbf{i} - 2\mathbf{j})$$
$$= 5\mathbf{i} + 12\mathbf{j}$$
$$|\mathbf{R}| = \sqrt{5^2 + 12^2}$$
$$= 13\,\text{N}$$

$$a = \frac{F}{m} = \frac{13}{5} = 2.6\,\text{m s}^{-2}$$

> To find the resultant you add both vectors in the **i** direction and then both vectors in the **j** direction. The resulting vector is the vector for the resultant. Pythagoras' theorem is then used to find the magnitude of the resultant.

(c) $\theta = \tan^{-1}\left(\dfrac{12}{5}\right) = 67.4°$ (1 d.p.)

Direction is 67.4° (1 d.p.) above the positive *x*-axis (or the unit **i** vector)

**5** Notice the velocity and distance are given, so the acceleration of the lift can be found before the tension in the cable is determined.

$$v^2 = u^2 + 2as \text{ with } u = 0\,\text{m s}^{-1}, v = 4\,\text{m s}^{-1}, s = 4\,\text{m}$$
$$4^2 = 0^2 + 2a \times 4$$

Giving   $a = 2\,\text{m s}^{-2}$

> As *t* is unknown, we need to use an equation of motion not containing *t*.

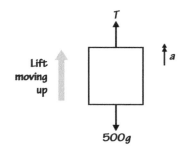

Applying Newton's 2nd law of motion to the lift in the vertical direction,
$$500a = T - 500g$$
$$500 \times 2 = T - 500 \times 9.8$$
$$T = 5900\,\text{N}$$

> Always draw a diagram showing the forces acting and the direction of the acceleration.